Lecture Notes in Mathematics

A collection of informal reports and seminars
Edited by A. Dold, Heidelberg and B. Eckmann, Zürich

133

Flemming Topsøe
Department of Mathematics, University of Copenhagen

Topology and Measure

Springer-Verlag
Berlin · Heidelberg · New York 1970

CONTENTS

PREFACE INTRODUCTION ACKNOWLEDGMENTS

Below we shall comment on the development which led to the results of the present volume (and of [26]).

It will be seen that our investigations took their starting point in the theory of weak convergence of (probability-) measures, and that the results on measure and integration theory to be found in part I emerged as a kind of "by-product".

During the inspiring lectures of professor P. Billingsley, I was for the first time presented to the theory of weak convergence. These lectures, which were based on the manuscript to the book [3], took place in 1964-65 while professor Billingsley visited the institute of math. statistics at the university of Copenhagen. The contact with Billingsley resulted, among other things, in the joint paper [4]. This paper was the starting point of the development leading to the results obtained in sections 12-17.

The academic year 1965-66 I spent at the statistical laboratory, the university of Cambridge, England. There I met professor K.R. Parthasarathy, who was at that time working on his book [17], dealing with weak convergence too. Once, professor Parthasarathy posed the following question and explained the significance of the problem: Let X be a Polish space, $\mathcal{A}$ a field of subsets of X generating the Borel σ-field and P, $(P_n)_{n\geq 1}$ probability measures on X such that $\lim P_n A = PA$ for all A in $\mathcal{A}$; is it then true, or under what additional assumptions is it true, that P_n converges weakly to P?

A solution to the problem of Parthasarathy was published in [24]. It was quite obvious that the reasoning had very little of "Polishness" in it and, with this motivation, we became interested in developing a theory of weak convergence of tight measures on arbitrary topological Hausdorff spaces. Of course, this had already been done by Varadarajan in 1961 and by his predecessors; however, despite the great achievements in this previous research, there were some points which seemed unsatisfactory to me (notably, the choice of σ-fields).

One of the first results we were able to obtain was that $\mathcal{M}_+(X;t)$ in its weak topology is a Hausdorff space for any Hausdorff space X (Theorem 11.2, (i)). This innocent looking result was in fact enough to convince us that we were working with a fruitful definition of weak convergence. Note, that the definition is based on semi-continuous functions and not on continuous functions. The definition and further study of the weak topology, was also influenced by the announcement [20] of L.Schwartz of a theory of Radon measures (= tight measures) on arbitrary Hausdorff spaces. The following remark cited from [20] was helpful: "All the properties of Radon measures given in Bourbaki extend here, provided continuous functions don't occur in their statement (but semi-continuous functions may occur)".

Clearly, the theory of weak convergence would be worthless, could we not establish suitable criteria for compactness. Led by certain observations (which are at present uninteresting) we came to the conviction, that Prohorovs condition of tightness would also be sufficient in the general case. I had a very specific idea as to how the hoped-for result could be proved but couldn't push it through. At that time (1968) I received affirmative solutions of the problem from P.A. Meyer and from L. Schwartz. Both used the idea - which never occured to me - of reducing the problem to the compact case.

The problem was settled, but for some reasons, perhaps stubbornness too, I was not fully satisfied and tried again - in vain - to push my own ideas through. Then a paper by J. Kisyński ([14]) was published which did just what we wanted to. Indeed, it turned out that with Kisyński's result at hand we could obtain necessary and sufficient conditions for compactness in $\mathcal{M}_+(X;t)$. However, the condition we arrived at (cf. Theorem 9.1) was rather complicated and we tried to derive simpler ones. We did nothave much luck with the tightness condition (see notes and remarks to section 9), and we started to look into the condition of τ-smoothness. Then one should work in the space $\mathcal{M}_+(X;\tau)$ rather than in the space $\mathcal{M}_+(X;t)$, and the compact sets in X did no longer play the dominant role, but was to some extent taken over by the closed sets. It was natural to examine once more the proof of Kisyński's theorem and see if one could axiomatize it so that it would cover the two cases. This was in fact quite easy to do (Theorem 2.2), and as a consequence we obtained the desired result on τ-smoothness and compactness (Theorem 9.2)

It turned out that Theorem 2.2 also contained the extension theorem in abstract measure theory known as Carathéodory's theorem (see [26]). Thus we have an instance of a result that can be used as the key <u>to topological measure theory as well as to abstract measure theory</u>. We found this point worth while exploring in its own right and, for that reason, we established an analogous result, Theorem 3.13, which is an extension theorem for integrals. This result contains a version of Daniells theorem. We do hope that these abstract results will turn out to be of some interest, also in the teaching of measure and integration theory. We are fully aware that we have contributed only a modest amount to the ideas in the proofs of Theorems 2.2 and 3.13. The main idea is still the brillant idea from 1914 due to C. Carathéodory (see [7]).

We have tried to round off the results on measure theory by inclusion of sections 4,5 and 6. Section 4 deals with finitely additive measures and integrals, and contains a version, due to A. Markoff, of the "Riesz representation theorem".

Provoked by a stimulating discussion with E.T. Kehlet, we included a section explaining, in the framework of the present theory, why it is possible, in locally compact spaces, to construct regular Borel-measures from certain set-functions only defined on the compact Baire-sets.

The main results in section 6 on construction of measures by approximation from outside or from inside are in fact vital for the solution of the original problem on compactness in the topology of weak convergence. Compactness problems in spaces of measures provided with other topologies than that of weak convergence can also be handled by appealing to the results of section 6 - one instance of this is demonstrated in the following section dealing with the vague topology, and yet another topology, much stronger than the ones already mentioned, is discussed in our paper [26].

It is probably true, that all small tricks employed in the first six sections can be found scattered in the huge literature of measure and integration theory. I have made no attempt to make reference to this literature - for one reason, I do not know it well enough, and, for another, it seems as if the main idea, viz. the idea to base the theory of measure on inner measure, has not been taken up previously. There is, however, at least one exception: As pointed out to us by S.D. Chatterji, the paper [21] from 1955 by Srinivasan works explicitely with inner measure. Also, it may be true that some of the results to be found in part I (notably Theorem 2.2) are contained in the paper [18] from 1951 by Pettis.

We have now described how the work was carried out and what the motivations were. In the text we have of course arranged the material in a more systematic order. There are two parts. Part II deals with weak convergence andpart I contains the material not directly connected with weak convergence. The main theme of part I is construction of measures and integrals. A section of notes and remarks is included at the end of the text.

The reader, who is not interested in going through the entire text may find the following suggestions for reading attractive: Sections 1,2, 3 (omitting all proofs), 5(again omitting proofs), 6 (only paying attention to Theorem 6.2), 8,9,10, and then, at last, the reader has to decide how much of sections 12-17 he wants to go through; this is perhaps best done by looking into sections 15-17 where the results of sections 12-14 are applied, but even if the applications are found to be worth while, the reader may find it unbearable to go through the proofs of sections 13 and 14. However, in order to understand what is going on, it is quite sufficient to acquire familiarity with the proofs of the much simpler Theorem 2 of Billingsley and Topsøe [4] and of Theorem 2 of Topsøe [22]. It can be said that once the problem in the more complicated set-up has been properly formulated (cf. section 12), it is a matter of routine to solve it, knowing the ideas of the above mentioned papers.

It follows from what is said above, that the persons I am mostly indebted to are professor Billingsley and professor Parthasarathy, and I want to express my sincere thanks.

PRELIMINARIES

P1. $\mathbb{R}$ denotes the reals and $\mathbb{R}_+$ the non-negative reals.

P2. Sometimes we find it convinient to call a non-empty class of subsets of a set X, a <u>paving</u> on X. A $(\cup f,\cap c)$-paving is a paving closed under finite unions and countable intersections. If in addition the empty set $\emptyset$ is a member of the paving, we speak of a $(\emptyset,\cup f,\cap c)$-paving. This method of notation is employed systematically through the text. The paving of all subsets of X is denoted $\mathcal{D}(X)$.

P3. A class $\mathcal{A} \subseteq \mathcal{D}(X)$ is <u>filtering to the left</u> or <u>filtering downwards</u> if, to any pair (A_1,A_2) of sets in $\mathcal{A}$ we can find $A \in \mathcal{A}$ such that $A \subseteq A_1 \cap A_2$. We write $\mathcal{A} \downarrow A_o$ if $\mathcal{A}$ is filtering to the left and $A_o = \bigcap\{A \mid A \in \mathcal{A}\}$. A class $\mathcal{F}$ of functions $f: X \to \mathbb{R}$ filters to the left if $f_1,f_2 \in \mathcal{F} \Rightarrow \exists f \in \mathcal{F} : f \leq \min(f_1,f_2)$. Filtering to the right (or upwards) is defined analogously.

P4. We shall work with nets rather than with filters. Our prefered labels for directed sets are D,E and I. Elements of a directed set denoted by the letter D[I] will always be denoted by the letter α[i]. Examples: $(x_\alpha)_{\alpha\in D}$, $(f_i)_{i\in I}$ or just (x_α), (f_i). We write $x_\alpha \in A$, eventually or just $x_\alpha \in A$, ev. if for some $\alpha_o \in D$ we have $x_\alpha \in A$ for all $\alpha \geq \alpha_o$. We write $x_\alpha \in A$, frequently or just $x_\alpha \in A$, freq. if for every $\alpha \in D$ we have $x_\beta \in A$ for some $\beta \geq \alpha$.

P5. A <u>notion of convergence</u> $\rightarrow$ on a set X is a class of pairs (ξ,x) where ξ is a net on X and x a point of X such that certain conditions are fulfilled. We write $x_\alpha \rightarrow x$ to indicate that the pair $((x_\alpha)_{\alpha\in D},x)$ is a member of the class. We require that the following conditions are fulfilled:

(i): $x_\alpha = x$ for all $\alpha \in D \Rightarrow x_\alpha \to_\tau x$,

(ii): $x_\alpha \to_\tau x \Rightarrow x_{\alpha_\beta} \to_\tau x$ for every subnet (x_{α_β}) of (x_α),

(iii): $x_\alpha \to_\tau x,\ x_\alpha \to_\tau y \Rightarrow x = y$,

(iv): If $(x_\alpha)_{\alpha\in D}$, x is such that every subnet of (x_α) contains a further subnet converging to x, then the net (x_α) itself converges to x.

P6. We shall assume that all our basic topological spaces are Hausdorff spaces. Thus every topological space denoted by the letter X is assumed to be a Hausdorff space. For a topological space X we denote by $\mathcal{F}(X)$, $\mathcal{K}(X)$, $\mathcal{G}(X)$ and $\mathcal{B}(X)$ the pavings on X of closed, compact, open and Borel sets, respectively. Sets denoted by the letters F,K,G are usually assumed without further mentioning to be closed, compact or open, respectively.

P7. A net (x_α) on a topological space X is said to be <u>compact</u> if every subnet has a further subnet which converges (or, equivalently, if every universal subnet of (x_α) converges). A subset A of X is called <u>net-compact</u> if every net on A has a convergent subnet (or, equivalently, if every universal net on A converges). In case X is a regular topological space, $A \subseteq X$ is net-compact if and only if A is relatively compact.

P8. Let X be a topological space and (F_α) a net on $\mathcal{F}(X)$. We define two sets F_* and F^*, both closed sets, by

$$F_* = \{x \mid \forall_{N(x)}\ N(x) \cap F_\alpha \neq \emptyset,\ \text{ev.}\},$$

$$F^* = \{x \mid \forall_{N(x)}\ N(x) \cap F_\alpha \neq \emptyset,\ \text{freq.}\}.$$

Here, N(x) denotes some neighbourhood of x. If $F_* = F^*$, we write $F_\alpha \to F_*$ and we say that F_α <u>converges in the notion of closed topological convergence</u> to F_*. This notion is indeed a notion of convergence on $\mathcal{F}(X)$, and it has the interesting property (Hausdorffs selection theorem) that every net on $\mathcal{F}(X)$ has a convergent subnet. The notion is topological if and only if X is locally compact.

P9. For X a topological space and f a function $X \to \hat{R}$ (say bounded) we define the lower semi-continuous envelope f_* and the upper semi-continuous envelope f^* of f by

$$f_* = \sup\{g \mid g \leq f,\ g \text{ l.s.c.}\},$$
$$f^* = \inf\{g \mid g \geq f,\ g \text{ u.s.c.}\}.$$

f_* is l.s.c., f^* is u.s.c., $f_* \leq f \leq f^*$ holds, and, furthermore, we have

$$D(f) = \{f_* < f^*\},$$

where D(f) denotes the set of discontinuity points of f. It follows that

$$D(f) = \bigcup_1^\infty \{f^* - f_* \geq 1/n\}$$

is a F_σ-set, in particular a Borel-set.

P10. Let X be a topological space. A class $\mathcal{A} \subseteq \mathcal{P}(X)$ is said to separate points T_2 if to any pair (x,y) of distinct points in X there exists $A \in \mathcal{A}$ such that $x \in \mathring{A}$ and $y \notin \bar{A}$. $\mathcal{A}$ is said to separate points and closed sets T_1 if $x \notin F$, $F \in \mathcal{F}(X) \Rightarrow x \in \mathring{A}$, $A \cap F = \emptyset$ for some $A \in \mathcal{A}$. $\mathcal{A}$ is said to separate points and closed sets T_2 if $x \notin F$, $F \in \mathcal{F}(X) \Rightarrow x \in \mathring{A}$, $\bar{A} \cap F = \emptyset$ for some $A \in \mathcal{A}$. If $\mathcal{A}$ separate points and closed sets T_1 and if X is regular then $\mathcal{A}$ separate points and closed sets T_2.

P11. Let X be an abstract set and $\mathcal{A}$ and $\mathcal{S}$ two classes of subsets of X such that: $\mathcal{A} \subseteq \mathcal{S}$, $\mathcal{A}$ is closed under finite intersections, the complement of every set in $\mathcal{S}$ is in $\mathcal{S}$, the union of two disjoint sets in $\mathcal{S}$ is in $\mathcal{S}$, and $S_1 \setminus S_2 \in \mathcal{S}$ whenever $S_1, S_2 \in \mathcal{S}$ and $S_1 \supseteq S_2$. Then $\alpha(\mathcal{A})$, the algebra spanned by $\mathcal{A}$, is contained in $\mathcal{S}$.

P12. By a set-function we mean a non-negative, possibly infinite valued function defined on a paving. Let β be a set-function defined on the paving $\mathcal{A}$. In the definitions below we only require the defining relations to hold when they make sense. β is monotone if $A_1 \subseteq A_2 \Rightarrow \beta A_1 \leq \beta A_2$.

β is <u>subadditive</u> if $\beta(A_1 \cup A_2) \leq \beta A_1 + \beta A_2$ holds. β is <u>additive</u> if $A_1 \cap A_2 = \emptyset \Rightarrow \beta(A_1 \cup A_2) = \beta A_1 + \beta A_2$. β is <u>modular</u> if $\emptyset \in \mathcal{A}$, if $\beta\emptyset = 0$, and if $\beta(A_1 \cup A_2) + \beta(A_1 \cap A_2) = \beta A_1 + \beta A_2$ holds. A monotone set function β defined on $\mathcal{A}$ is <u>σ-smooth</u> [<u>τ-smooth</u>] <u>with respect to the paving</u> $\mathcal{K}$ if, for any countable [arbitrary] subclass $\mathcal{K}^*$ of $\mathcal{K}$ which filters downwards to a set A_o in $\mathcal{A}$ ($\mathcal{K}^* \downarrow A_o$), we have

$$\beta A_o = \inf\{\beta A \mid A \supseteq K^* \text{ for some } K^* \in \mathcal{K}^*\},$$

provided the right hand side in this equation is finite. If $\emptyset \in \mathcal{A}$ and if we only require the last relation to hold when $A_o = \emptyset$, then we obtain the definition of set-functions which are <u>σ-smooth at $\emptyset$</u> [<u>τ-smooth at $\emptyset$</u>] <u>w.r.t.</u> $\mathcal{K}$. If $\mathcal{K} = \mathcal{A}$ in the last definitions, we call the set-function <u>σ-smooth</u>, <u>τ-smooth</u>, <u>σ-smooth at $\emptyset$</u>, or <u>τ-smooth at $\emptyset$</u>, respectively. β is <u>regular w.r.t. the paving</u> $\mathcal{K}$ if $\mathcal{K} \subseteq \mathcal{A}$ and if

$$\beta A = \sup\{\beta K \mid K \subseteq A,\ K \in \mathcal{K}\}$$

holds. β is <u>tight</u> if β is finite and if, whenever $A_1 \supseteq A_2$, the relation

$$\sup\{\beta A \mid A \subseteq A_1 \setminus A_2\} = \beta A_1 - \beta A_2$$

holds. β is a <u>content</u> if $\mathcal{A}$ is a $(\emptyset, \cup f, \cap f)$ paving and if β is finite, monotone, subadditive and additive.

P13. Let X be a topological space. By $\mathcal{M}_+(X)$ we denote the space of all non-negative totally finite measures defined on $\mathcal{B}(X)$. $\mu \in \mathcal{M}_+(X)$ is <u>regular</u> if μ is regular w.r.t. the paving $\mathcal{F}(X)$, and μ is <u>tight</u>, or a <u>Radon measure</u>, if μ is regular w.r.t. the paving $\mathcal{K}(X)$. $\mu \in \mathcal{M}_+(X)$ is <u>τ-smooth</u> if

$$\mu\Big(\bigcap_{F \in \mathcal{F}} F\Big) = \inf_{F \in \mathcal{F}} \mu F$$

holds for any family $\mathcal{F}$ of closed sets filtering to the left. $\mathcal{M}_+(X;r)$, $\mathcal{M}_+(X;\tau)$, $\mathcal{M}_+(X;t)$, $\mathcal{M}_+(X;r,\tau)$ denote the sets of regular, τ-smooth, tight, and regular τ-smooth measures in $\mathcal{M}_+(X)$, respectively.

P14. If $A \in \mathcal{B}(X)$ and $\mu \in \mathcal{M}_+(X;t)$ then $\mu|A$, the restriction of μ to A, is a measure in $\mathcal{M}_+(X,t)$ too. An analogous statement holds for $\mathcal{M}_+(X;r,\tau)$.

P15. If $\mu \in \mathcal{M}_+(X;t)$ then $\mu \in \mathcal{M}_+(X;r,\tau)$. If $\mu \in \mathcal{M}_+(X;\tau)$ and if $\mathcal{F}$ is a uniformly bounded family of u.s.c. functions $X \to \hat{\mathbb{R}}$ filtering to the left then we have

$$\mu(\inf_{f \in \mathcal{F}} f) = \inf_{f \in \mathcal{F}} \mu(f).$$

This is easily proved by applying the inequality

$$\frac{1}{k} \sum_1^k \mu(\{g \geq \frac{\nu}{k}\}) \leq \int g d\mu \leq \frac{\mu X}{k} + \frac{1}{k} \sum_1^k \mu(\{g \geq \frac{\nu}{k}\})$$

valid for any $k \geq 1$ and for any measurable function $g: X \to \hat{\mathbb{R}}$ with $0 < g < 1$.

P16. If every closed subset of X is a G_δ-set than every measure (in $\mathcal{M}_+(X)$) is regular.

If X is fully Lindelöf (i.e. every family of open sets has a countable subfamily with the same union), then every measure is τ-smooth.

If X can be metrized with a complete metric, or if X is locally compact, then every τ-smooth measure is tight.

If X is regular and μ τ-smooth then we have

$$\mu G_0 = \sup\{\mu G \mid \bar{G} \subseteq G_0\}; \; G_0 \in \mathcal{G}(X),$$

$$\mu F_0 = \inf\{\mu F \mid \mathring{F} \supseteq F_0\}; \; F_0 \in \mathcal{F}(X).$$

In particular, it follows that μ is regular.

If X is completely regular and μ τ-smooth then

$$\mu G_0 = \sup\{\mu G \mid G \subseteq G_0, \; \mu(\partial G) = 0\}.$$

$$= \sup\{\mu F \mid F \subseteq G_0, \; \mu(\partial F) = 0\}; \quad G_0 \in \mathcal{G}(X).$$

$$\mu F_0 = \inf\{\mu F \mid F \supseteq F_0, \; \mu(\partial F) = 0\}.$$

$$= \inf\{\mu G \mid G \supseteq F_0, \; \mu(\partial G) = 0\}; \; F_0 \in \mathcal{F}(X).$$

P17. A subset F of X is called <u>support</u> for $\mu \in \mathcal{M}_+(X)$, and we write $F = \mathrm{supp}(\mu)$, if $F \in \mathcal{F}(X)$, if $\mu F = \mu X$ and if $F' \in \mathcal{F}(X)$, $\mu F' = \mu X$ $\Rightarrow F' \supseteq F$. $\mathrm{supp}(\mu)$ is uniquely determined.

Every τ-smooth measure has a support.

P18. A function $f: X \to \hat{R}$ is called a μ-continuity function if f is measurable (w.r.t. $\mathcal{B}(X)$) and if $\mu(D(f)) = 0$. The set $A \in \mathcal{B}(X)$ is a μ-continuity set if $\mu(\partial A) = 0$.

P19. If $\mathcal{A}$ is a $(\cup f, \cap f)$-paving on X separating points T_2 and if μ_1 and μ_2 are two tight measures such that

$$\mu_1(\mathring{A}) \leq \mu_2(\bar{A}); \quad A \in \mathcal{A}$$

then $\mu_1 \leq \mu_2$. If, in addition $\mu_1 X = \mu_2 X$, then $\mu_1 = \mu_2$.

If X is regular and $\mathcal{A}$ a $(\cup f)$-paving on X separating points and closed sets T_2 and if μ_1 and μ_2 are measures in $\mathcal{M}_+(X;\tau)$ such that

$$\mu_1(\mathring{A}) \leq \mu_2(\bar{A}); \quad A \in \mathcal{A}$$

then $\mu_1 \leq \mu_2$. If, in addition, $\mu_1 X = \mu_2 X$, then $\mu_1 = \mu_2$.

PART I

1. Measure and integral, definitions.

By a **measure** we shall understand a triple $(X, \mathcal{B}, \mu)$ where X is a set, $\mathcal{B}$ a σ-field of subsets of X and μ a countably additive set-function $\mathcal{B} \to \hat{R}_+ \cup \{\infty\}$ with $\mu(\emptyset) = 0$. When there is no danger of ambiguity we speak of "the measure μ". μ is complete if every subset of a null-set is measurable (i.e. belongs to $\mathcal{B}$). With a given measure μ we associate two set functions μ_*, the inner measure, and μ^*, the outer measure, both defined on the class of all subsets of X:

$$\mu_*(A) = \sup\{\mu B : A \supseteq B \in \mathcal{B}\} \quad ; A \subseteq X,$$

$$\mu^*(A) = \inf\{\mu B : A \subseteq B \in \mathcal{B}\} \quad ; A \subseteq X.$$

Let $(X, \mathcal{B}, \mu)$ be a measure, denote by $\mathcal{E}$ the class of all functions $X \to \hat{R}_+ \cup \{\infty\}$ measurable with respect to $\mathcal{B}$, and denote by $\mathcal{H}$ the class of all simple functions in $\mathcal{E}$ (i.e. functions of the form $\Sigma_1^n \alpha_i \cdot 1_{A_i}$ with the α's in $\hat{R}_+$ and the A's in $\mathcal{B}$).

For functions in $\mathcal{E}$ we define the integral w.r.t. μ as follows: Firstly, for $h = \Sigma \alpha_i 1_{A_i}$ in $\mathcal{H}$, we define $\int h d\mu$ by

$$\int h d\mu = \Sigma \alpha_i \mu(A_i),$$

and, for f in $\mathcal{E}$, we define $\int f d\mu$ by

$$\int f d\mu = \sup\{\int h d\mu : f \geq h \in \mathcal{H}\}.$$

We shall now define what we mean in general by an integral. The integrals we shall consider will be positive (as is the case with the measures), and in the construction of integrals we shall only consider positive functions. In the class of all functions $X \to \hat{R}_+ \cup \{\infty\}$ we have the lattice operations expressed by the signs $\vee$ and $\wedge$, we have addition and multiplication with positive scalars and, furthermore, we find it convenient to introduce the operation $\setminus$ defined by

$$f_1 \setminus f_2 = \begin{cases} f_1 - f_1 \wedge f_2 & \text{for } f_2 < \infty \\ 0 & \text{for } f_2 = \infty \end{cases}$$

In other words, $f_1 \setminus f_2$ is the smallest function for which the decomposition

$$f_1 = f_1 \wedge f_2 + f_1 \setminus f_2$$

holds. The following identities

$$f_1 \setminus f_2 = f_1 - f_1 \wedge f_2 = f_1 \vee f_2 - f_2 = (f_1 - f_2)^+$$

are valid, provided we interprete $\infty - \infty$ as 0.

Consider a class $\mathcal{E}$ of functions $X \to \mathring{R}_+ \cup \{\infty\}$ containing the zero-function 0, and such that, for any choice of functions f_n in $\mathcal{E}$, all functions

$$\bigvee_1^\infty f_n \ , \ \bigwedge_1^\infty f_n \ , \ f_1 + f_2 \ , \ f_1 \setminus f_2 \ , \ \alpha f \ (\text{for } \alpha \in \mathring{R}_+)$$

are members of $\mathcal{E}$. Such a class we shall call a $(0, \vee c, \wedge c, \setminus)$ convex cone of functions.

By an <u>integral</u> we shall understand a triple $(X, \mathcal{E}, I)$ where X is a set, where $\mathcal{E}$ is a $(0, \vee c, \wedge c, \setminus)$ convex cone of functions $X \to \mathring{R}_+ \cup \{\infty\}$, and where $I: \mathcal{E} \to \mathring{R}_+ \cup \{\infty\}$ is positively homogeneous and additive and satisfies the condition that, for any increasing sequence $(f_n)_{n \geq 1}$ of functions in $\mathcal{E}$, the relation

$$I(\bigvee_1^\infty f_n) = \sup_{n \geq 1} I(f_n)$$

holds. The condition that I be positively homogeneous and additive means that the relation

$$I(\Sigma\, \alpha_i f_i) = \Sigma\, \alpha_i I(f_i)$$

holds for any finite sum (α's in $\mathring{R}_+$, and f's in $\mathcal{E}$).

The integral I is <u>complete</u> if every non-negative function g dominated by a function f in $\mathcal{E}$ with $I(f) = 0$ is itself a member of $\mathcal{E}$.

If $(X, \mathcal{E}, I)$ is an integral, we shall usually denote by $\mathcal{B}$ the paving on X consisting of all sets $A \subseteq X$ for which the indicator-function 1_A is a member of $\mathcal{E}$, and we shall denote by μ the set-function on $\mathcal{B}$ defined by

$$\mu(A) = I(1_A); \quad A \in \mathcal{B} .$$

If the constant function 1 is in $\mathcal{E}$ then (and only then) $(X, \mathcal{B}, \mu)$ is a measure, the underlying measure. In that case it is easy to see that every function $X \to \hat{R}_+ \cup \{\infty\}$ measurable with respect to $\mathcal{B}$ is a member of $\mathcal{E}$ and that $I(f) = \int f d\mu$ for every such function. We shall call the integral $(X, \mathcal{E}, I)$ a full integral if $1 \in \mathcal{E}$ and if the class $\mathcal{E}$ coincides with the class of functions measurable w.r.t. the underlying σ-field $\mathcal{B}$.

The inner and outer integrals are defined by the formulas

$$I_*(f) = \sup\{I(h): f \geq h \in \mathcal{E} \}; \quad f \geq 0,$$

$$I^*(f) = \inf\{I(h): f \leq h \in \mathcal{E} \}; \quad f \geq 0.$$

2. Basic result on construction of a measure. Let $\mathcal{K}$ denote a $(\emptyset, \cup f, \cap f)$-paving on the abstract set X; for the most important results $\mathcal{K}$ will in fact be a $(\emptyset, \cup f, \cap c)$-paving, or even a $(\emptyset, \cup f, \cap a)$-paving. Examples are $\mathcal{K}(X)$ or $\mathcal{F}(X)$ if X is a topological space, or we might consider the paving of compact Baire sets or the paving of closed Baire sets.

By $\mathcal{F}(\mathcal{K})$ and $\mathcal{G}(\mathcal{K})$ we denote the two pavings on X defined by

$$\mathcal{F}(\mathcal{K}) = \{F \mid K \cap F \in \mathcal{K} \quad \forall K \in \mathcal{K} \},$$

$$\mathcal{G}(\mathcal{K}) = \{G \mid K \setminus G \in \mathcal{K} \quad \forall K \in \mathcal{K} \}.$$

$\mathcal{F}(\mathcal{K})$ is a $(\emptyset, X, \cup f, \cap f)$-paving containing $\mathcal{K}$. If $\mathcal{K}$ is, say a $(\emptyset, \cup f, \cap c)$-paving then $\mathcal{F}(\mathcal{K})$ is a $(\emptyset, X, \cup f, \cap c)$-paving. $\mathcal{G}(\mathcal{K})$ consists of the complements of the sets in $\mathcal{F}(\mathcal{K})$. By $\mathcal{B}(\mathcal{K})$ we denote the σ-field spanned by $\mathcal{G}(\mathcal{K})$. If one wishes, one could call the sets in $\mathcal{F}(\mathcal{K})$, $\mathcal{G}(\mathcal{K})$, and $\mathcal{B}(\mathcal{K})$ the "$\mathcal{K}$-closed", the "$\mathcal{K}$-open", and the "$\mathcal{K}$-Borel" sets, respectively.

From Topsøe [26] we borrow the following results:

Lemma 2.1. Let λ be a tight set-function on the $(\emptyset,\cup f,\cap f)$-paving $\mathcal{K}$. Then λ is monotone and modular, in particular a content. If λ is σ-smooth at $\emptyset$ [τ-smooth at $\emptyset$] then λ is σ-smooth [τ-smooth]. If $\mathcal{K}$ is semi-compact [compact] then λ is σ-smooth [τ-smooth].

Theorem 2.2. Let $\mathcal{K}$ be a $(\emptyset,\cup f,\cap c)$-paving on X, and assume that $\lambda:\mathcal{K}\to\hat{R}_+$ is tight and σ-smooth at $\emptyset$. Define $\mu_*: \mathcal{D}(X)\to\hat{R}_+\cup\{\infty\}$ by

$$\mu_*A = \sup\{\lambda K \mid K \subseteq A\};\ A \subseteq X.$$

Consider the class

$$\mathcal{B} = \{A \subseteq X \mid \mu_*K = \mu_*(K\cap A) + \mu_*(K\setminus A)\ \forall K\},$$

and denote by μ the restriction of μ_* to $\mathcal{B}$.

Then $(X, \mathcal{B}, \mu)$ is a complete measure and the σ-field $\mathcal{B}$ contains $\mathcal{B}(\mathcal{K})$. Furthermore, $(X, \mathcal{B}, \mu)$ is the largest extension of λ to a measure regular w.r.t. $\mathcal{K}$.

This result is basic for much of the material to follow. As a simple consequence we mention here that one can, in a natural way, define the "Radon part" of a measure on a topological space X. Let $\mu \in \mathcal{M}_+(X)$ (we may in fact replace the condition $\mu(X) < \infty$ by $\mu(K) < \infty\ \forall K \in \mathcal{K}(X)$, and also, we need only assume that μ be finitely additive). Then the restriction of μ to $\mathcal{K}(X)$ is a tight content, σ-smooth at $\emptyset$. Thus the formula

$$\mu_t(A) = \sup_{K\subseteq A} \mu(K);\quad A \in \mathcal{B}(X)$$

defines a measure $\mu_t \in \mathcal{M}_+(X;t)$. We call μ_t the Radon part of μ. μ_t is characterized by the decomposition $\mu = \mu_t+\nu$ where $\mu_t \in \mathcal{M}_+(X;t)$, $\nu \in \mathcal{M}_+(X)$, and $\nu K = 0\ \forall K \in \mathcal{K}(X)$.

Lemma 2.3. Assume that, in the set-up of Theorem 2.2., λ is in fact τ-smooth at $\emptyset$. Then the measure μ constructed in Theorem 2.2 will be τ-smooth w.r.t. the paving $\mathcal{F}(\mathcal{K})$ of $\mathcal{K}$-closed sets.

Proof. Assume for simplicity that $\mathcal{K}$ is a $(\emptyset,\cup f,\cap a)$-paving. Let $\mathcal{F}$ be a subclass of $\mathcal{F}(\mathcal{K})$ such that $\mathcal{F}\downarrow F_0$ say, and such that μF_1 is finite for some $F_1 \in \mathcal{F}$. To $\varepsilon > 0$ we choose $K \subseteq F_1$ such that $\mu_*(F_1\setminus K)\leq\varepsilon$.

Since the class

$$\{F \cap K \mid F \in \mathcal{F}, F \subseteq F_1\}$$

is a subclass of $\mathcal{K}$ filtering to the left to $F_0 \cap K$, we can by Lemma 2.1 find a set $F \in \mathcal{F}$ with $F \subseteq F_1$ and $\mu(F \cap K) \leq \mu(F_0 \cap K) + \varepsilon$. Then

$$\mu F = \mu(F \cap K) + \mu(F \setminus K) \leq \mu(F_0 \cap K) + \varepsilon + \mu(F_1 \setminus K)$$

$$\leq \mu F_0 + 2\varepsilon.$$

□

Let λ be a content on the $(\emptyset, \cup f, \cap f)$-paving $\mathcal{K}$, and denote, as in Theorem 2.2, by μ_* the set-function on $\mathcal{D}(X)$ defined from λ by regularity w.r.t. $\mathcal{K}$. We call λ <u>locally finite</u> provided we, to any $K \in \mathcal{K}$, can find a set $G \in \mathcal{G}(\mathcal{K})$ with $G \supseteq K$ and $\mu_*(G) < \infty$. If, furthermore, it is true that

$$\lambda K = \inf\{\mu_*(G) \mid G \supseteq K,\ G \in \mathcal{G}(\mathcal{K})\}$$

holds for all $K \in \mathcal{K}$, then λ is said to be <u>semi-regular</u>.

<u>Lemma</u> 2.4. <u>Let</u> λ <u>be a content on a</u> $(\emptyset, \cup f, \cap f)$ <u>paving</u> $\mathcal{K}$.

(i) <u>If</u> λ <u>is locally finite and tight then</u> λ <u>is semi-regular</u>.

(ii) <u>If</u> $\mathcal{G}(\mathcal{K})$ <u>separate the sets in</u> $\mathcal{K}$ <u>and if</u> λ <u>is semi-regular then</u> λ <u>is tight</u>.

<u>Proof</u>. (i): Instead of basing the proof on Theorem 2.2 - which would demand further assumptions - we appeal to Theorem 4.1 of section 4. Denote by μ the finitely additive measure constructed there. If $K \in \mathcal{K}$, consider a $\mathcal{K}$-open set $G \supseteq K$ with $\mu G < \infty$. To a given $\varepsilon > 0$ we choose $K' \subseteq G \setminus K$ such that $\mu(G \setminus K) \leq \mu(K') + \varepsilon$. From the equality

$$\mu(G \setminus K') + \mu(K') = \mu(G \setminus K) + \mu(K)$$

we see that $\mu(G \setminus K') \leq \mu(K) + \varepsilon$. Since $G \setminus K'$ is $\mathcal{K}$-open and contains K, we conclude that λ is semi-regular.

(ii): Consider sets K_1, K_2 (in $\mathcal{K}$) with $K_1 \subseteq K_2$. Let ε be positive. Choose $G \supseteq K_1$ such that $\mu_*(G) \leq \lambda K_1 + \varepsilon$. Since $\mathcal{G}(\mathcal{K})$ separate the sets in $\mathcal{K}$, we can find $G' \supseteq K_1$ and $G'' \supseteq K_2 \setminus G$ such that $G' \cap G'' = \emptyset$. The set $K_2 \setminus G'$ is in $\mathcal{K}$ and is contained in $K_2 \setminus K_1$. Furthermore, we have

$$\begin{aligned}\lambda(K_2\setminus G')+\lambda K_1 &\geq \lambda(K_2\setminus G')+\mu_*(G)-\varepsilon\\ &\geq \lambda(K_2\setminus G')+\lambda(K_2\setminus G'')-\varepsilon\\ &\geq \lambda((K_2\setminus G')\cup(K_2\setminus G''))-\varepsilon\\ &= \lambda(K_2)-\varepsilon.\end{aligned}$$

This argument shows that $\mu_*(K_2\setminus K_1) \geq \lambda K_2-\lambda K_1$. The reverse inequality follows from the additivity of λ. Thus λ is tight. ▯

3. Basic result on construction of an integral. In this section we shall prove a result analogous to Theorem 2.2 but based on a class of functions instead on a class of sets.

Throughout the section, X is an abstract set and $\mathcal{K}$ a convex cone of non-negative finite valued functions defined on X, closed under finite suprema and countable infima, containing the zero-function 0, and satisfying Stones condition that $k\wedge 1 \in \mathcal{K}$ for every $k \in \mathcal{K}$. In short, $\mathcal{K}$ is a $(0,\vee f,\wedge c)$ convex cone of functions $X\to\mathring{R}_+$ satisfying Stones condition.

Furthermore, we assume throughout the section that λ: $\mathcal{K}\to\mathring{R}_+$ is a non-negative finite valued map defined on $\mathcal{K}$ which is σ-smooth at 0 and tight. Tightness means that the relation

$$\sup\{\lambda(k) \mid k \leq k_1-k_2\} = \lambda(k_1)-\lambda(k_2)$$

holds whenever $k_1 \geq k_2$ holds (functions denoted by the letter "k" are always assumed to be members of $\mathcal{K}$).

By $\mathcal{U}_+(\mathcal{K})$ we denote the class of functions f: $X\to\mathring{R}_+\cup\{\infty\}$ for which $k\wedge f$ is a member of $\mathcal{K}$ for every k in $\mathcal{K}$, and by $\mathcal{B}(\mathcal{K})$ we denote the smallest σ-field on X for which all functions in $\mathcal{U}_+(\mathcal{K})$ are measurable. One could call the functions in $\mathcal{U}_+(\mathcal{K})$ the non-negative "$\mathcal{K}$-upper semi-continuous" functions.

It is our goal to extend λ to a full integral $(X, \mathcal{E}, I)$, regular w.r.t. $\mathcal{K}$, and such that $\mathcal{E}$ contains $\mathcal{U}_+(\mathcal{K})$. To do this, we first define a map I_*, mapping the class of all functions $X\to\mathring{R}_+\cup\{\infty\}$ into $\mathring{R}_+\cup\{\infty\}$:

$$I_*(f) = \sup\{\lambda(k) \mid k \le f\};\ f \ge 0.$$

Then we define $\mathcal{E}$ by

$$\mathcal{E} = \{f \mid I_* k = I_*(k \wedge f) + I_*(k \setminus f)\ \forall k\}$$

(here, as below, all functions are functions $X \to \hat{R}_+ \cup \{\infty\}$). The restriction of I_* to $\mathcal{E}$ is the desired integral, but it will take some time before we can see this.

The reader who is also interested in the finitely additive theory should notice that a good deal of what we are going to say below remains true in the more general setup where $\mathcal{K}$ is only assumed to be a $(0, \vee f, \wedge f)$ convex cone satisfying Stones condition and where $\lambda: \mathcal{K} \to \hat{R}_+$ is only assumed to be tight.

We note that I_* is an extension of λ and that the tightness of λ can be expressed in terms of I_* by the relation

(3.1) $$I_*(k_1 - k_2) = I_*(k_1) - I_*(k_2);\quad k_1 \ge k_2.$$

Lemma 3.1. λ is monotone, $\lambda(0) = 0$, λ is positively homogeneous and additive, λ is modular and λ is σ-smooth.

Proof. The additivity of λ follows from the considerations

$$\lambda(k_1) = I_*((k_1 + k_2) - k_2) = I_*(k_1 + k_2) - I_*(k_2)$$
$$= \lambda(k_1 + k_2) - \lambda(k_2).$$

The modularity of λ follows from the considerations

$$I_*(k_1 \vee k_2) - I_*(k_2) = I_*(k_1 \vee k_2 - k_2)$$
$$= I_*(k_1 - k_1 \wedge k_2) = I_*(k_1) - I_*(k_1 \wedge k_2).$$

To prove that λ is σ-smooth, assume that $k_n \downarrow k_o$ and let $\varepsilon > 0$ be given. Choose a sequence (ε_n) of positive numbers such that $\Sigma_1^\infty \varepsilon_n < \varepsilon$ holds. Choose a sequence (k_n') such that $k_n' \le k_n - k_o$; $n \ge 1$ and such that

$$\lambda(k_n') + \lambda(k_o) \ge \lambda(k_n) - \varepsilon_n;\ n \ge 1.$$

We claim that, for $n \ge 1$, the inequality

(3.2) $$\lambda(k_n) \le \lambda(k_o) + \lambda(k_1' \wedge \ldots \wedge k_n') + \Sigma_1^n \varepsilon_\nu$$

holds. This is clear for $n = 1$. Assume that (3.2) holds for a specific n and let us establish (3.2) with n replaced by n+1. We have:

$$
\begin{aligned}
&\lambda(k_0) + \lambda(k_1' \wedge \cdots \wedge k_{n+1}') \\
&= \lambda(k_0) + \lambda(k_1' \wedge \cdots \wedge k_n') + \lambda(k_{n+1}') - \lambda\big((k_1' \wedge \cdots \wedge k_n') \vee k_{n+1}'\big) \\
&\geq \lambda(k_n) - \textstyle\sum_1^n \varepsilon_\nu + \lambda(k_{n+1}') - \lambda\big((k_1' \wedge \cdots \wedge k_n') \vee k_{n+1}'\big) \\
&\geq \lambda(k_{n+1}) - \textstyle\sum_1^{n+1} \varepsilon_\nu + I_*(k_n - k_0) - I_*\big((k_1' \wedge \cdots \wedge k_n') \vee k_{n+1}'\big) \\
&\geq \lambda(k_{n+1}) - \textstyle\sum_1^{n+1} \varepsilon_\nu ,
\end{aligned}
$$

and have thus established (3.2) with n replaced by n+1. By induction we see that (3.2) holds for all $n \geq 1$. Since $k_1' \wedge \ldots \wedge k_n' \downarrow 0$ and since λ is σ-smooth at 0, we conclude from (3.2) that n can be chosen so that $\lambda(k_n) \leq \lambda(k_0) + \varepsilon$ holds. This argument shows that λ is σ-smooth. The remaining parts of the proof are left to the reader. ▯

Lemma 3.2. $\mathcal{U}_+(\mathcal{K})$ is contained in $\mathcal{E}$; if $f \in \mathcal{E}$ and if $a \in \dot{R}_+$ then $a \cdot f \in \mathcal{E}$.

I_* is positively homogeneous, and for arbitrary functions f and g we have

$$I_*(f+g) \geq I_*(f) + I_*(g).$$

If $f \in \mathcal{E}$, then we have, for any function g:

$$I_*(g) = I_*(g \wedge f) + I_*(g \setminus f). \tag{3.3}$$

Again, if $f \in \mathcal{E}$, then we have, for any function g:

$$I_*(f+g) = I_*(f) + I_*(g). \tag{3.4}$$

Proof. The first statements are simple to prove. To prove (3.3), we argue as follows:

$$
\begin{aligned}
I_*(g) &= \sup\{\lambda(k) \mid k \leq g\} \\
&= \sup\{I_*(k \wedge f) + I_*(k \setminus f) \mid k \leq g\} \\
&\leq I_*(g \wedge f) + I_*(g \setminus f).
\end{aligned}
$$

(3.4) follows from (3.3) since we have

$$
\begin{aligned}
I_*(f+g) &= I_*((f+g) \wedge f) + I_*((f+g) \setminus f) \\
&= I_*(f) + I_*((f+g) \setminus f) \\
&\leq I_*(f) + I_*(g).
\end{aligned}
$$

▯

Lemma 3.3. *Let* $(f_n)_{n\geq 1}$ *be a sequence of functions in* $\mathcal{E}$, *let* $k \in \mathcal{K}$ *and let* ε *be positive. Choose two sequences* $(k'_n)_{n\geq 1}$ *and* $(k''_n)_{n\geq 1}$ *of functions in* $\mathcal{K}$ *such that*

$$k'_n \leq k \wedge f_n ; \quad n \geq 1,$$

$$k''_n \leq k \setminus f_n ; \quad n \geq 1,$$

and such that

$$\lambda(k) \leq \lambda(k'_n) + \lambda(k''_n) + \varepsilon \cdot 2^{-n} ; \quad n \geq 1$$

hold.

Then the two sets of inequalities

(3.5) $$\lambda(\bigvee_1^n k'_i) + \lambda(\bigwedge_1^n k''_i) \geq \lambda(k) - \Sigma_1^n \, \varepsilon \cdot 2^{-i} ; \quad n \geq 1 ,$$

and

(3.6) $$\lambda(\bigwedge_1^n k'_i) + \lambda(\bigvee_1^n k''_i) \geq \lambda(k) - \Sigma_1^n \, \varepsilon \cdot 2^{-i} ; \quad n \geq 1$$

hold.

Proof. The proofs of (3.5) and (3.6) are somewhat similar; therefore, we shall only give the proof of (3.5). For n = 1, (3.5) does hold. To establish the general validity of (3.5) we need only prove that, for $n \geq 1$, the inequality

$$\lambda(\bigvee_1^{n+1} k'_i) + \lambda(\bigwedge_1^{n+1} k''_i) \geq \lambda(\bigvee_1^n k'_i) + \lambda(\bigwedge_1^n k''_i) - \varepsilon \cdot 2^{-(n+1)}$$

holds. By modularity, this inequality can be rewritten in the form

$$\lambda(k'_{n+1}) + \lambda(k''_{n+1}) \geq \lambda(k'_{n+1} \wedge \bigvee_1^n k'_i) + \lambda(k''_{n+1} \vee \bigwedge_1^n k''_i) - \varepsilon \cdot 2^{-(n+1)}$$

By assumption, it is enough to prove the inequality

$$\lambda\left[(k'_{n+1} \wedge \bigvee_1^n k'_i) + (k''_{n+1} \vee \bigwedge_1^n k''_i)\right] \leq \lambda(k)$$

But this inequality is obviously fulfilled since we have

$$\begin{aligned} &(k'_{n+1} \wedge \bigvee_1^n k'_i) + (k''_{n+1} \vee \bigwedge_1^n k''_i) \\ &\leq (k \wedge f_{n+1}) \wedge \bigvee_1^n (k \wedge f_i) + (k \setminus f_{n+1}) \vee \bigwedge_1^n (k \setminus f_i) \\ &= k . \end{aligned}$$

▯

If we apply the σ-smoothness of λ in (3.5), we find that

$$(3.7) \qquad \lambda(\bigvee_1^n k_i') + \lambda(\bigwedge_1^\infty k_i'') \geq \lambda(k)-2\varepsilon \quad \text{ev.}$$

holds. It follows, that

$$I_*(k \wedge \bigvee_1^\infty f_n) + I_*(k \setminus \bigvee_1^\infty f_n) \geq \lambda(k) - 2\varepsilon$$

and, due to the freedom of choice of k and ε, we conclude that $\bigvee_1^\infty f_n \in \mathcal{E}$. In the same way, it follows from (3.6) that $\bigwedge_1^\infty f_n \in \mathcal{E}$.

From (3.7) it also follows that we have

$$\lambda(k) \leq \lim_n I_*(\bigvee_1^n f_i) + I_*(k \setminus \bigvee_1^\infty f_i) \quad ; k \in \mathcal{K} .$$

Then:

$$\begin{aligned} I_*(\bigvee_1^\infty f_i) &= \sup\{\lambda(k) \mid k \leq \bigvee_1^\infty f_i\} \\ &\leq \sup\{\lim_n I_*(\bigvee_1^n f_i) + 0 \mid k \leq \bigvee_1^\infty f_i\} \\ &= \lim_n I_*(\bigvee_1^n f_i). \end{aligned}$$

Let us collect the results proved above:

Lemma 3.4. *If $(f_n)_{n\geq 1}$ is a sequence of functions in $\mathcal{E}$ then $\bigvee_1^\infty f_n$ and $\bigwedge_1^\infty f_n$ are functions in $\mathcal{E}$; moreover, we have*

$$I_*(\bigvee_1^\infty f_n) = \lim_{n\to\infty} I_*(\bigvee_1^n f_i).$$

So far, all the steps we have taken have been more or less straightforward, taking the corresponding steps in the proof of Theorem 2.2 into account.

Lemma 3.5. *If f and g are in $\mathcal{E}$, then so is f+g, and, if we further assume that $I_*(f \wedge g)$ is finite, then the function $f \setminus g$ is in $\mathcal{E}$ too.*

Proof. For $k \in \mathcal{K}$ we have

$$\begin{aligned} I_*(k) &= I_*(k \wedge f) + I_*(k \setminus f) \\ &= I_*(k \wedge f) + I_*((k \setminus f) \wedge g) + I_*((k \setminus f) \setminus g) \\ &= I_*((k \wedge f) + (k \setminus f) \wedge g) + I_*((k \setminus f) \setminus g) \\ &= I_*(k \wedge (f+g)) + I_*(k \setminus (f+g)), \end{aligned}$$

and it follows that $f+g \in \mathcal{E}$.

Assume now that $I_*(f \wedge g)$ is finite. We may and do assume that $f \geq g$ holds. We leave it to the reader to conclude from the identities

$$k + g = k\setminus(f\setminus g) + (k+g)\wedge f,$$

and

$$(k+g)\wedge f = k\wedge(f\setminus g) + g,$$

and from the information already gained at this stage, that $f\setminus g \in \mathcal{E}$. ▯

We now introduce the class

$$\mathcal{B} = \{ A \subseteq X \mid 1_A \in \mathcal{E} \}$$

and want to find the relation between $\mathcal{B}$ and $\mathcal{E}$.

Lemma 3.6. *$\mathcal{B}$ is a σ-algebra, and if $f\colon X \to \bar{R}_+ \cup \{\infty\}$ is $\mathcal{B}$-measurable then f belongs to $\mathcal{E}$* .

Proof. To prove that $\mathcal{B}$ is a σ-algebra we need only prove that $A \in \mathcal{B} \Rightarrow \complement A \in \mathcal{B}$. Assume then that $A \in \mathcal{B}$. Put $B = \complement A$ and consider a function k in $\mathcal{K}$. We have

$$\begin{aligned} &I_*(k\wedge 1_B) + I_*(k\setminus 1_B) \\ &= I_*(k\setminus 1_A - k\setminus 1) + I_*(k\wedge 1_A + k\setminus 1); \end{aligned}$$

now it follows from Stones condition and from Lemma 3.5 that $k\setminus 1 = k\setminus(k\wedge 1) \in \mathcal{E}$, hence, by equation (3.4), we can continue our calculations as follows:

$$\begin{aligned} \ldots &= I_*(k\setminus 1_A) - I_*(k\setminus 1) + I_*(k\wedge 1_A) + I_*(k\setminus 1) \\ &= I_*(k), \end{aligned}$$

and we conclude that $B \in \mathcal{B}$.

To prove the second half of the lemma we need only prove that the non-negative simple functions (w.r.t. $\mathcal{B}$) are in $\mathcal{E}$. Assume then that f can be written in the form

$$f = \Sigma_1^n \alpha_i \cdot 1_{A_i}$$

where $\alpha_i \in \bar{R}_+$, $A_i \in \mathcal{B}$; $i=1,\ldots,n$. We may and do assume that the A's are pairwise disjoint. Then f can also be written in the form

$$f = \bigvee_1^n (\alpha_i \cdot 1_{A_i});$$

thus $f \in \mathcal{E}$. ▯

We now aim at proving that every function in $\mathcal{E}$ is $\mathcal{B}$-measurable.

<u>Lemma</u> 3.7. <u>If</u> $f\backslash a \in \mathcal{E}$ <u>for all positive constants</u> a <u>then</u> f <u>is</u> $\mathcal{B}$-<u>measurable</u>.

<u>Proof.</u> Let $a > 0$ and put $A = \{x \in X \mid f(x) > a\}$. Then

$$1_A = 1 \wedge \big(\infty \cdot (f\backslash a)\big),$$

hence $1_A \in \mathcal{E}$. ▯

<u>Corollary</u> 3.8. <u>If</u> $f \in \mathcal{E}$ <u>and</u> $k \in \mathcal{K}$ <u>then</u> $k \wedge f$ <u>is</u> $\mathcal{B}$-<u>measurable</u>.

This is a corollary to Lemma 3.7 and Lemma 3.5.

<u>Lemma</u> 3.9. <u>Let</u> $f \in \mathcal{E}$, $k \in \mathcal{K}$ <u>and let</u> a <u>and</u> b <u>be two real constants of which</u> a <u>is positive</u>. <u>Then the two sets</u>

$$\{k<a\} \cup \{f<b\}$$

<u>and</u>

$$\{k<a\} \cup \{f \geq b\}$$

<u>are both members of</u> $\mathcal{B}$.

<u>Proof</u>. We may assume that b is positive too. Put $k' = (a^{-1}b)k$. Then

$$\{k<a\} \cup \{f<b\} = \{(k' \wedge f) < b\},$$

and this set is in $\mathcal{B}$, due to Corollary 3.8. From the rearrangement

$$\{k<a\} \cup \{f \geq b\} = \complement\Big[\big(\{k<a\} \cup \{f<b\}\big) \setminus \{k<a\}\Big]$$

we then see that $\{k<a\} \cup \{f \geq b\} \in \mathcal{B}$. ▯

<u>Lemma</u> 3.10. <u>If</u> f <u>and</u> g <u>are in</u> $\mathcal{E}$, <u>if</u> g <u>is finite, and if</u> $f \geq g$ <u>then</u> $f-g \in \mathcal{E}$.

<u>Proof</u>. Fix $k \in \mathcal{K}$ and $a > 0$. We have

$$\{k \wedge (f-g) < a\} = \{k<a\} \cup \bigcup_r (\{f<r\} \cap \{g \geq r-a\})$$

$$= \bigcup_r [(\{k<a\} \cup \{f<r\}) \cap (\{k<a\} \cup \{g \geq r-a\})]$$

where the unions over r are to be taken over all rational r. By Lemma 3.9 it follows that $\{k \wedge (f-g) < a\} \in \mathcal{B}$. Thus $k \wedge (f-g)$ is $\mathcal{B}$-measurable. By Lemma 3.6, $k \wedge (f-g) \in \mathcal{E}$, and then it follows by Lemma 3.2 that we have

$$I_*(k) = I_*\big(k \wedge (f-g)\big) + I_*\big(k \backslash (f-g)\big).$$

Since k was an arbitrary function in $\mathcal{K}$, it follows that $f-g \in \mathcal{E}$. ▯

Finally, we get

Lemma 3.11 If f is in $\mathcal{E}$ then f is $\mathcal{B}$-measurable.

Proof. For a > 0 the formula $f \backslash a = f - f \wedge a$ and Lemma 3.10 shows that $f \backslash a \in \mathcal{E}$. By Lemma 3.7 it follows that f is $\mathcal{B}$-measurable. ▯

We have now seen that $(X, \mathcal{E}, I)$ is a full integral (I denoting the restriction of I_* to $\mathcal{E}$). Let the underlying measure be $(X, \mathcal{B}, \mu)$. We want to say something about the regularity of μ. For this purpose we define the trace of $\mathcal{K}$, denoted by $tr(\mathcal{K})$, as the paving on X consisting of all sets K of the form

$$K = \{k \geq a\}$$

where $k \in \mathcal{K}$ and where a is strictly positive.

Lemma 3.12. $tr(\mathcal{K})$ is a $(\emptyset, \cup f, \cap c)$ paving on X and the underlying measure μ is regular w.r.t. $tr(\mathcal{K})$.

Proof. The proof that $tr(\mathcal{K})$ is a $(\emptyset, \cup f, \cap c)$ paving is left to the reader. Now, let $B \in \mathcal{B}$ and $\varepsilon > 0$ be given. Choose $k \in \mathcal{K}$ such that $k \leq 1_B$ and such that

$$\mu(B) = I_*(1_B) \leq I_*(k) + \varepsilon.$$

Then choose the positive number a such that

$$I_*(k \wedge a) \leq \varepsilon.$$

The set $K = \{k \geq a\}$ is contained in B, is a member of $tr(\mathcal{K})$, and we have

$$\begin{aligned} \mu(B) &\leq I_*(k) + \varepsilon \\ &= I_*(k \backslash a) + I_*(k \wedge a) + \varepsilon \\ &\leq I_*(1_{\{k \geq a\}}) + 2\varepsilon \\ &= \mu(K) + 2\varepsilon. \end{aligned}$$

▯

Let us state in full detail the result we have proved by now

Theorem 3.13. Let X be an abstract space, $\mathcal{K}$ a $(0, \vee f, \wedge c)$ convex cone of functions $X \to \hat{R}_+$ satisfying Stones condition, and let $\lambda: \mathcal{K} \to \hat{R}_+$

be tight and σ-smooth at 0. Define I_* and $\mathcal{E}$ by

$$I_*(f) = \sup\{\lambda(k) \mid k \le f\},$$

$$\mathcal{E} = \{f \mid I_*(k) = I_*(k \wedge f) + I_*(k \setminus f) \ \forall k\},$$

and denote by I the restriction of I_* to $\mathcal{E}$.

Then $(X, \mathcal{E}, I)$ is a full and complete integral, regular w.r.t. $\mathcal{K}$; furthermore, the underlying σ-field contains $\mathcal{B}(\mathcal{K})$, and the underlying measure is regular w.r.t. $\mathrm{tr}(\mathcal{K})$.

Finally, we can characterize $(X, \mathcal{E}, I)$ as the largest extension of $(X, \mathcal{K}, \lambda)$ to an integral, regular w.r.t. $\mathcal{K}$.

The final statement and the completeness of I have not been mentioned above but they are both easy to establish.

If X is a topological space, we can apply Theorem 3.13 either to the class $\mathcal{K}$ of all bounded upper semi-continuous functions $:X \to \hat{R}_+$ or to the class of functions in $\mathcal{K}$ with compact support. In the latter case, the condition that λ be σ-smooth at 0 is superfluous due to Dinis lemma.

We shall now indicate how the Daniell extension theorem follows from Theorem 3.13. We consider an abstract set X provided with a convex cone $\mathcal{H}$ of functions $X \to \hat{R}_+$ such that $h_1 \wedge h_2$, $h_1 \vee h_2$ and $h_1 \setminus h_2$ are in $\mathcal{H}$ whenever h_1 and h_2 are in $\mathcal{H}$ and such that $h \wedge 1 \in \mathcal{H}$ for all $h \in \mathcal{H}$. In short, $\mathcal{H}$ is a $(0, \vee f, \wedge f, \setminus)$ convex cone of functions $X \to \hat{R}_+$ satisfying Stones condition. Furthermore, there is given a positively homogeneous and additive functional $I: \mathcal{H} \to \hat{R}_+ \cup \{\infty\}$ which is σ-smooth at 0 $\left(h_n \downarrow 0,\ I(h_1) < \infty \Rightarrow I(h_n) \downarrow 0\right)$. We wish to extend I to an integral. For that purpose define $\mathcal{K}$ and $\lambda: \mathcal{K} \to \hat{R}_+$ by

$$\mathcal{K} = \{k \mid h_n \downarrow k \text{ for some sequence in } \mathcal{H} \text{ with } I(h_1) < \infty\},$$

$$\lambda(k) = \inf\{I(h) \mid k \le h \in \mathcal{H}\};\ k \in \mathcal{K} .$$

It is fairly straightforward to check that $\mathcal{K}, \lambda$ satisfy all the hypothesis of Theorem 3.13 (for example, we may proceed as follows: (i): $\mathcal{K}$ has the required structure, (ii): λ is σ-smooth at 0, (iii): $h_n \downarrow k$,

$I(h_1)<\infty \Rightarrow I(h_n)\downarrow\lambda(k)$, (iv): $\lambda(k_1+k_2) = \lambda(k_1) + \lambda(k_2)$, (v): λ is tight). Theorem 3.13 gives us an extension $(X, \mathcal{E}, \hat{I})$ of λ. Clearly, $\mathcal{H} \subseteq \mathcal{E}$ (since $\mathcal{H} \subseteq \mathcal{U}_+(\mathcal{K})$) and $\hat{I}(h) = I(h)$ for all h in $\mathcal{H}$ with $I(h) < \infty$. If we assume that I has no atom h with $I(h) = \infty$ (i.e. assume that $I(h) = \infty$ implies $0<I(h')<\infty$ for some $h'\leq h$) then $\hat{I}(h) = I(h)$ holds for all $h \in \mathcal{H}$.

We can apply the Daniell extension theorem established above to prove that any measure $(X, \mathcal{B}, \mu)$ can be extended to an integral. To see this, let $\mathcal{H}$ denote the class of non-negative simple functions and let $I: \mathcal{K} \to \hat{R}_+\cup\{\infty\}$ be defined by

$$I(h) = \int h d\mu; \quad h \in \mathcal{H} .$$

Then the desired result follows in case μ has no atoms of infinite measure ($\mu A = \infty \Rightarrow 0<\mu B<\infty$ for some $B \subseteq A$); a separate argument will allow us to remove this restriction. In other words, we have proved, as is well known, that the mapping $f\to\int f d\mu$, where f is measurable w.r.t. $\mathcal{B}$, is an integral.

4. Finitely additive theory.

By a finitely additive measure we shall understand a triple $(X, \mathcal{A}, \mu)$, where X is a set, $\mathcal{A}$ an algebra of subsets of X and $\mu: \mathcal{A} \to \hat{R}_+\cup\{\infty\}$ an additive set-function with $\mu(\emptyset) = 0$.

Theorem 4.1. Let $\mathcal{K}$ be a $(\emptyset,\cup f,\cap f)$ paving on X and $\lambda: \mathcal{K} \to \hat{R}_+$ a tight content. Define μ_* by

$$\mu_*(A) = \sup\{\lambda K \mid K \subseteq A\}; \ A \in \mathcal{D}(X) ,$$

and define $\mathcal{A}$ by

$$\mathcal{A} = \{A \mid \mu_*(K) = \mu_*(K\cap A) + \mu_*(K\setminus A) \ \forall K \in \mathcal{K}\}.$$

Denote by μ the restriction of μ_* to $\mathcal{A}$.

Then $(X, \mathcal{A}, \mu)$ is a finitely additive measure and the algebra $\mathcal{A}$ contains the algebra spanned by $\mathcal{G}(\mathcal{K})$. Furthermore, $(X, \mathcal{A}, \mu)$ is the largest extension of $(X, \mathcal{A}, \lambda)$ to a finitely additive measure regular w.r.t. $\mathcal{K}$ and this extension is complete.

This result is simpler to prove than Theorem 2.2 and we shall not pause to give the details.

Assume now that $(X, \mathcal{A}, \mu)$ is a finitely additive measure and that $\mu(X)$ is finite. By a simple function we shall as usual understand a function of the form

$$f = \Sigma_1^n \alpha_i \cdot 1_{A_i}$$

where the α's are in $\hat{R}_+$ and the A's in $\mathcal{A}$. The integral of the simple function f w.r.t. μ is defined by

$$\int f d\mu = \Sigma_1^n \alpha_i \cdot \mu(A_i).$$

Let now $f: X \to \hat{R}_+$ be a bounded function; we shall say that f is <u>integrable</u> (w.r.t. μ) if, for every positive ε, we can find two simple functions g and h such that $g \leq f \leq h$ and such that $\int(h-g)d\mu < \varepsilon$; if this condition is fulfilled we define

$$\int f d\mu = \sup\{\int g d\mu \mid g \leq f, \ g \text{ simple}\}.$$

An arbitrary function $f: X \to \hat{R}_+$ is said to be <u>integrable</u> if $f \wedge n$ is integrable for every n and if

$$\sup\{\int g d\mu \mid g \leq f, \ g \text{ simple}\}$$

is finite; if these conditions are fulfilled then $\int f d\mu$ is defined as the supremum above.

<u>Lemma</u> 4.2. <u>Let</u> $(X, \mathcal{A}, \mu)$ <u>be a finitely additive measure with</u> $\mu(X)<\infty$ <u>and denote by</u> $\mathcal{E}$ <u>the class of functions</u> $f: X \to \hat{R}_+$ <u>integrable w.r.t.</u> μ. <u>Then</u> $\mathcal{E}$ <u>is a</u> $(0, \vee f, \wedge f, \setminus)$ <u>convex cone of functions</u>.

<u>Furthermore, any bounded function</u> f <u>such that the sets</u> $\{f \geq a\}$, <u>for any</u> $a>0$ <u>are members of</u> $\mathcal{A}$ <u>is integrable</u>.

The proof of the first part of the lemma is very tedious. The proof of the second part involves a standard argument. Details are left to the reader.

By a <u>finitely additive integral</u> we shall understand a triple $(X, \mathcal{E}, I)$ where X is a set, where $\mathcal{E}$ is a $(0, \vee f, \wedge f, \setminus)$ convex cone of functions

$X \to \hat{R}_+$, and where $I: \mathcal{E} \to \hat{R}_+$ is finite and positively homogeneous and additive.

If $(X, \mathcal{E}, I)$ is a finitely additive integral, we shall usually denote by $\mathcal{A}$ the paving on X consisting of all sets A for which $1_A \in \mathcal{E}$, and we shall denote by μ the set-function on $\mathcal{A}$ defined by

$$\mu(A) = I(1_A); \quad A \in \mathcal{A} .$$

If $1 \in \mathcal{E}$ then $(X, \mathcal{A}, \mu)$ is a finitely additive measure, the _underlying finitely additive measure._ $(X, \mathcal{E}, I)$ is said to be _full_ if $1 \in \mathcal{E}$ and if $\mathcal{E}$ consists of precisely those functions that are integrable w.r.t. μ.

Lemma 4.3. _If_ $(X, \mathcal{E}, I)$ _is a full finitely additive integral with underlying finitely additive measure_ μ _and if_

$$I(f) = \sup_n I(f \wedge n)$$

holds for all $f \in \mathcal{E}$, _then we have_

$$I(f) = \int f d\mu; \quad f \in \mathcal{E} .$$

The simple proof is left to the reader.

Theorem 4.4. _Let_ $\mathcal{K}$ _be a_ $(0, \vee f, \wedge f)$ _convex cone of functions_ $X \to \hat{R}_+$ _satisfying Stones condition, and let_ $\lambda: \mathcal{K} \to \hat{R}_+$ _be tight. Define_ I_* _and_ $\mathcal{E}$ _by_

$$I_*(f) = \sup\{\lambda(k) \mid k \leq f\}; \quad f: X \to \hat{R}_+,$$

$$\mathcal{E} = \{f \mid I_*(f) < \infty \text{ and } I_*(k) = I_*(k \wedge f) + I_*(k \setminus f) \ \forall k \in \mathcal{K} \},$$

and denote by I _the restriction of_ I_* _to_ $\mathcal{E}$. _Assume that_ $I(1) < \infty$. _Define_ $\mathcal{A}$ _and_ μ _by_

$$\mathcal{A} = \{A \subseteq X \mid 1_A \in \mathcal{E} \},$$

$$\mu(A) = I(1_A); \quad A \in \mathcal{A} .$$

Then $(X, \mathcal{A}, \mu)$ _is a finitely additive measure. Assume now, furthermore, that every function in_ $\mathcal{K}$ _is bounded and that_ $tr(\mathcal{K}) \subseteq \mathcal{A}$ _(in particular we have that every function in_ $\mathcal{K}$ _is integrable w.r.t._ μ_)._

Then $(X, \mathcal{E}, I)$ _is a full and complete finitely additive integral,_ regular w.r.t. $\mathcal{K}$, and μ, the underlying finitely additive measure is regular w.r.t. tr($\mathcal{K}$) and we have

$$I(f) = \int f d\mu$$

for all $f \in \mathcal{E}$. Furthermore, $\mathcal{E}$ contains all bounded functions in $\mathcal{U}_+(\mathcal{K})$.

Finally, we can characterize $(X, \mathcal{E}, I)$ as the largest extension of $(X, \mathcal{K}, \lambda)$ to a finitely additive integral regular w.r.t. $\mathcal{K}$.

Proof. To begin with, one adapts lemmas 3.1-3.6 to the present situation. Among other things, we will then see that $(X, \mathcal{A}, \mu)$ is a finitely additive measure and that every simple function (w.r.t. $\mathcal{A}$) is in $\mathcal{E}$. Denote by $\mathcal{E}^{\mu}$ the class of integrable functions w.r.t. μ.

Lemma 4.5. If $f: X \to \hat{R}_+$ is such, that for any $\varepsilon > 0$ there exist two functions g and h in $\mathcal{E}$, such that $g \leq f \leq h$ and such that $I_*(h-g) < \varepsilon$, then f is in $\mathcal{E}$.

We leave the simple proof of this lemma to the reader.

Lemma 4.6. $\mathcal{E}^{\mu} \subseteq \mathcal{E}$.

Proof. Assume that $f \in \mathcal{E}^{\mu}$. If f is bounded then it follows by Lemma 4.5, the definition of $\mathcal{E}^{\mu}$ and by the fact that the simple functions are in $\mathcal{E}$, that $f \in \mathcal{E}$. If f is unbounded, it follows from the just said and from the assumption that each function in $\mathcal{K}$ is a bounded function in $\mathcal{E}^{\mu}$, that $k \wedge f \in \mathcal{E}$ for every $k \in \mathcal{K}$; hence $f \in \mathcal{E}$. $\square$

Lemma 4.7. Let h be a bounded function in $\mathcal{E}^{\mu}$. Then every function in $\mathcal{E}$ dominated by h is in $\mathcal{E}^{\mu}$.

Proof. Let $f \leq h$ and $f \in \mathcal{E}$. Then $I_*(h) = I_*(f) + I_*(h-f)$ holds, and therefore, to a given $\varepsilon > 0$ we can find $k_1 \leq f$ and $k_2 \leq h-f$ such that

$$\lambda(k_1) + \lambda(k_2) \geq I_*(h)-\varepsilon$$

holds. Then

$$I_*\big((h-k_2)-k_1\big) = I_*(h) - I_*(k_1+k_2)$$

$$\leq \varepsilon$$

follows. Now note that the functions k_1 and $h-k_2$ are bounded functions in $\mathcal{E}^{\mu}$ and that $k_1 \leq f \leq h-k_2$ holds. Since it is easy to see that for bounded functions in $\mathcal{E}^{\mu}$, $I_*(\cdot)$ and $\int \cdot d\mu$ agree, we also have

$$\int [(h-k_2)-k_1] d\mu \leq \varepsilon .$$

Even though the functions k_1 and $h-k_2$ need not be simple functions it is of course easy to deduce from these considerations that $f \in \mathcal{E}^{\mu}$. ▯

Lemma 4.8. $\mathcal{E} \subseteq \mathcal{E}^{\mu}$.

Proof. If $f \in \mathcal{E}$, it follows by Lemma 4.7 that $f \wedge n \in \mathcal{E}^{\mu}$ for all $n \geq 1$. Since we also have

$$\sup\{\int g d\mu \mid g \leq f,\ g \text{ simpel}\} = \sup\{I_*(g) \mid g \leq f,\ g \text{ simpel}\}$$
$$\leq I_*(f) < \infty ,$$

it follows that $f \in \mathcal{E}^{\mu}$. ▯

We have now established the most important parts of Theorem 4.4.

The validity of the formula $I(f) = \int f d\mu;\ f \in \mathcal{E}$ follows from Lemma 4.3 (since the functions in $\mathcal{K}$ are bounded).

That μ is regular w.r.t. $tr(\mathcal{K})$ follows from the assumption $tr(\mathcal{K}) \subseteq \mathcal{A}$ and from Lemma 3.12 (the fact that $\inf\{I(k \wedge a) \mid a > 0\} = 0$ is now obtained from the finiteness assumption $I(1) < \infty$).

We have now completed the proof of Theorem 4.4. ▯

We shall use Theorem 4.4 to derive one of the many versions of the "Riesz representation theorem". We start with some remarks on closed Baire sets. X is an arbitrary topological space. By $\mathcal{F}_0(X)$ we denote the paving of closed Baire sets i.e. the paving on X of all sets of the form $F = h^{-1}(\{0\})$ for some continuous function $h: X \to \hat{R}$. In this definition we may assume that h is a mapping $X \to [0,1]$. $\mathcal{F}_0(X)$ is a $(\emptyset, X, \cup f, \cap c)$ paving on X. If A is a closed subset of $\hat{R}$, and if $h: X \to \hat{R}$ is continuous then $h^{-1}(A) \in \mathcal{F}_0(X)$. By $C_+(X)$ we denote the class of non-negative continous and bounded functions on X. By $\mathcal{U}^{!}_{+}(X)$ we denote the $(\wedge c)$-closure of $C_+(X)$ i.e. $\mathcal{U}_+^!(X)$ consists of all functions that can be written as a countable infima of functions in $C_+(X)$. By $\mathcal{L}_+^!(X)$ we denote the (vc)-closure of $C_+(X)$. The paving $\mathcal{F}_0(X)$ can also be descrcibed as the trace of $\mathcal{U}_+^!(X)$. Let us collect some of these facts in

Lemma 4.9. The paving $\mathcal{F}_0(X)$ of closed Baire sets on a topological space X is a $(\emptyset, X, \cup f, \cap c)$ paving, and we have

$$tr(\mathcal{U}_+^!(X)) = \mathcal{F}_0(X).$$

We also need the following version of the "in-between theorem":

Lemma 4.10. Let k be a function in $\mathcal{U}_+^!(X)$, let g be a function in $\mathcal{L}_+^!(X)$ and assume that $k \leq g$ holds. Then there exists a function $f \in C_+(X)$ such that $k \leq f \leq g$.

The proof of this lemma follows a well known idea due to Hausdorff (cf. [10]) - the reader may also find it convenient to consult Hing Tong, [11].

Theorem 4.11. Let X be an arbitrary topological space and let there be given a positively homogenous and additive mapping $I: C_+(X) \to \hat{R}_+$. Then there exists a finitely additive measure μ defined on the algebra spanned by $\mathcal{F}_0(X)$, regular w.r.t. $\mathcal{F}_0(X)$, and such that

$$I(f) = \int f d\mu$$

for all $f \in C_+(X)$. μ is uniquely determined.

Proof. Denote by λ the set function on $\mathcal{U}_+^!(X)$ defined by

$$\lambda(k) = \inf\{I(h) \mid h \geq k,\ h \in C_+(X)\};\ k \in \mathcal{U}_+^!(X).$$

By an elementary reasoning resembling that in the proof of Lemma 2, (i) of Topsøe [26] we find that if λ is additive then λ is a tight content. By the in-between theorem (Lemma 4.10) it is easy to see that λ is additive, thus λ is a tight content. We are now in position to apply Theorem 4.4, and with only a few further considerations we find that the finitely additive measure μ obtain from Theorem 4.4 is regular w.r.t. $\mathcal{F}_0(X)$ and satisfies $\int h d\mu = I(h)$; $h \in C_+(X)$. To prove uniqueness, assume now, conversely, that some finitely additive measure μ defined on the algebra spanned by $\mathcal{F}_o(X)$ satisfies these conditions. We shall prove that for any set $F \in \mathcal{F}_o(X)$, the equality

$$\mu(F) = \inf\{I(h) \mid h \geq 1_F\}$$

holds. The "$\leq$-part" of this equality is trivially fulfilled. To prove the remaining part, let $\varepsilon > 0$ be given and choose $F' \subseteq \complement F$ with $F' \in \mathcal{F}_o(X)$ and such that

$$\mu(F') + \mu(F) \geq \mu(X)-\varepsilon$$

holds. We may apply the in-between theorem (Lemma 4.10) to the functions 1_F and 1_G where $G = \complement F'$. Thus there exists a continuous h with $0 \leq h \leq 1$, vanishing on F' and assuming the value 1 on F. For this function we have

$$I(h) = \int h d\mu \leq \mu(\complement F') \leq \mu(F)+\varepsilon$$

which implies the desired result. ▮

5. <u>From "Baire" measures to "Borel" measures, an abstract approach.</u>

Our basic theorem on construction of measures (Theorem 2.2) shows that one can bild a measure on the Borel field of a topological space by taking as starting point a suitable set-function defined on the paving of compact sets. From the classical theory (see Halmos, [9]) we know that from some points of view, it is appropriate with a more modest starting point viz. a set function defined only on the compact Baire sets. The paving $\mathcal{K}_o(X)$ of compact Baire sets on the topological space X (say locally compact) is a $(\emptyset,\cup f,\cap c)$ paving which is not only semi-compact but, indeed, compact; therefore, it seems natural to close $\mathcal{K}_o(X)$ under the operations $(\emptyset,\cup f,\cap a)$. If one does this, then we arrive at the paving $\mathcal{K}(X)$ of compact sets (assuming X locally compact). This is the reason why the classical approach works.

We shall now generalize the above considerations and take as starting point three objects X, $\mathcal{K}_o$ and λ_o, where X is an abstract set, where

$\mathcal{K}_o$ is a $(\emptyset,\cup f,\cap f)$-paving on X (if the reader wishes, he may assume that $\mathcal{K}_o$ is a $(\emptyset,\cup f,\cap c)$ paving), and where $\lambda_o : \mathcal{K}_o \to \hat{R}_+$ is tight and τ-smooth at $\emptyset$. Had we followed the introductery remarks more closely, then we would have assumed that $\mathcal{K}_o$ were compact, however, the present set-up is more general (see Lemma 2.1) and equally natural.

By $\mathcal{K}$ we shall denote the $(\emptyset,\cup f,\cap a)$-closure of $\mathcal{K}_o$, which is the same as the $\cap a$-closure of $\mathcal{K}_o$; in other words, $\mathcal{K}$ consists of all sets that can be written as an intersection (non-empty) of sets in $\mathcal{K}_o$. We wish to construct a measure μ extending λ_o and such that μ is regular and τ-smooth w.r.t. $\mathcal{K}$. Denoting by λ the restriction of μ to $\mathcal{K}$, we see that the only hope is the set function defined by

$$\lambda K = \inf\{\lambda_o(K_o) \mid K_o \supseteq K\};\ K \in \mathcal{K} .$$

Here, as below, we shall use the convention to denote sets in $\mathcal{K}_o$ by K_o, K_{o1}, K_{o2} etc. and sets in $\mathcal{K}$ by K, K_1, K_2 etc.

The problem is to prove that λ is tight and τ-smooth. This is carried out in several steps.

(i) λ <u>is an extension of</u> λ_o, λ <u>is monotone, subadditive and additive</u>.

<u>Proof</u>. Only the additivity of λ requires comment. Assume that $K_1 \cap K_2 = \emptyset$ and let $\varepsilon > 0$ be given. Choose $K_o \supseteq K_1 \cup K_2$ such that $\lambda_o(K_o) \leq \lambda(K_1 \cup K_2)+\varepsilon$. Consider the class

$$\{K_{o1} \cap K_{o2} \mid K_{o1} \supseteq K_1,\ K_{o2} \supseteq K_2,\ K_{o1} \cup K_{o2} \subseteq K_o\}.$$

Since this subclass of $\mathcal{K}_o$ filters to the left to $\emptyset$, we can find K_{o1} and K_{o2} of the specified type such that $\lambda_o(K_{o1} \cap K_{o2}) \leq \varepsilon$. Then

$$\begin{aligned} \lambda K_1 + \lambda K_2 &\leq \lambda_o(K_{o1}) + \lambda_o(K_{o2}) \\ &= \lambda_o(K_{o1} \cup K_{o2}) + \lambda(K_{o1} \cap K_{o2}) \\ &\leq \lambda(K_1 \cup K_2)+2\varepsilon . \end{aligned}$$

▯

(ii) $\mathcal{K}_o^* \subseteq \mathcal{K}_o$, $\mathcal{K}_o^* \downarrow K \Rightarrow \lambda(K) = \inf\{\lambda_o(K_o^*) \mid K_o^* \in \mathcal{K}_o^*\}$.

Proof. Let $\varepsilon > 0$ be given. Choose $K_o \supseteq K$ such that $\lambda_o(K_o) \leq \lambda(K)+\varepsilon$. Since

$$\{K_o^* \cup K_o \mid K_o^* \in \mathcal{K}_o^*\} \downarrow K_o$$

we can find K_o^* with $\lambda_o(K^*{}_o \cup K_o) \leq \lambda_o(K_o)+\varepsilon$. Then

$$\lambda_o(K_o^*) \leq \lambda_o(K_o^* \cup K_o) \leq \lambda_o(K_o)+\varepsilon \leq \lambda(K)+2\varepsilon.$$

▯

(iii) $K_1 \subseteq K_2,\ K_2 \in \mathcal{K}_o \Rightarrow \underset{\varepsilon>0}{\forall}\ \underset{K_o \subseteq K_2 \setminus K_1}{\exists}\ \lambda(K_o)+\lambda(K_1) \geq \lambda(K_2)-\varepsilon$

Proof. Choose, to the given $\varepsilon > 0$, a set K_{o1} with $K_1 \subseteq K_{o1} \subseteq K_2$ such that $\lambda_o(K_{o1}) \leq \lambda(K_1)+\varepsilon/2$. By tightness of λ_o we can find $K_o \subseteq K_2 \setminus K_{o1}$ ($\subseteq K_2 \setminus K_1$) such that $\lambda_o(K_o)+\lambda_o(K_{o1}) \geq \lambda(K_2)-\varepsilon/2$. Then

$$\begin{aligned}\lambda(K_o) + \lambda(K_1) &\geq \lambda_o(K_o) + \lambda_o(K_{o1})-\varepsilon/2\\ &\geq \lambda_o(K_2)-\varepsilon\\ &= \lambda(K_2)-\varepsilon.\end{aligned}$$

▯

(iv) λ <u>is tight</u>.

Proof. Consider K_1 and K_2 with $K_1 \subseteq K_2$. Let $\varepsilon > 0$ be given. Choose $K_{o2} \supseteq K_2$ such that

$$\lambda_o(K_{o2}) \leq \lambda(K_2)+\varepsilon.$$

Choose, according to (iii), $K_o \subseteq K_{o2} \setminus K_1$ such that

$$\lambda(K_o) + \lambda(K_1) \geq \lambda(K_{o2})-\varepsilon.$$

Put $K = K_2 \cap K_o$. The class of sets of the form $K_o' \cap K_o$ with $K_2 \subseteq K_o' \subseteq K_{o2}$ (and $K_o' \in \mathcal{K}_o$) filters to the left to the set K. According to (ii), we can then choose K_o' with $K_2 \subseteq K_o' \subseteq K_{o2}$ such that

$$\lambda_o(K_o'\cap K_o) \leq \lambda(K)+\varepsilon.$$

Denoting by μ_{o*} the set-function defined in the usual way from λ_o by regularity w.r.t. $\mathcal{K}_o$ we find:

$$\begin{aligned}\lambda_o(K_o)-\lambda_o(K_o'\cap K_o) &= \mu_{o*}(K_o\setminus K_o') \\ &\leq \mu_{o*}(K_{o2}\setminus K_o') \\ &= \lambda_o(K_{o2})-\lambda_o(K_o') \\ &\leq \lambda_o(K_{o2})-\lambda(K_2) \\ &\leq \varepsilon.\end{aligned}$$

It follows, that

$$\begin{aligned}\lambda(K) + \lambda(K_1) &\geq \lambda_o(K_o'\cap K_o) + \lambda(K_1)-\varepsilon \\ &\geq \lambda_o(K_o) + \lambda(K_1)-2\varepsilon \\ &\geq \lambda(K_{o2})-3\varepsilon \\ &\geq \lambda(K_2)-3\varepsilon.\end{aligned}$$

Since $K \subseteq K_2\setminus K_1$ it follows by the above reasoning that (with the usual notation)

$$\mu_*(K_2\setminus K_1) + \lambda K_1 \geq \lambda K_2$$

holds. By additivity, the reverse inequality also holds. ▯

(v) λ __is__ τ-__smooth at__ $\emptyset$.

__Proof__. Assume that $\mathcal{K}^*\downarrow\emptyset$ where $\mathcal{K}^* \subseteq \mathcal{K}$. Then the class

$$\{K_o \mid K_o \supseteq K^* \text{ for some } K^* \in \mathcal{K}^*\}$$

filters downwards to $\emptyset$ and we can, to a given $\varepsilon > 0$, find K_o and K^* with $K_o \supseteq K^*$ and $\lambda_o(K_o) \leq \varepsilon$. $\lambda(K^*) \leq \varepsilon$ follows. ▯

(vi) λ __is__ τ-__smooth__.

__Proof__. According to (iv) and (v) this follows from Lemma 2.1. ▯

We have solved our problem, but find it natural to ask one more question: If we use λ_o to define a measure μ_o (intuitively, a "Baire" measure) and λ to define a measure μ (intuitively, a "Borel" measure), can we then be sure that μ is an extension of μ_o? The answer is affirmative. Let us collect the information in

Theorem 5.1. Let $\mathcal{K}_o$ be a $(\emptyset,\cup f,\cap f)$-paving on X and λ_o a tight and τ-smooth content on $\mathcal{K}_o$.

Denote by $\mathcal{K}$ the $(\emptyset,\cup f,\cap a)$-closure of $\mathcal{K}_o$ and define λ on $\mathcal{K}$ by

$$\lambda(K) = \inf\{\lambda_o(K_o) \mid K_o \supseteq K\};\ K \in \mathcal{K} .$$

Then λ is a tight and τ-smooth content on $\mathcal{K}$ which extends λ_o.

Assume now, furthermore, that $\mathcal{K}_o$ is a $(\emptyset,\cup f,\cap c)$-paving. Denote by $(X, \mathcal{B}_o,\mu_o)$ the largest extension of λ_o to a measure regular w.r.t. $\mathcal{K}_o$, and denote by $(X, \mathcal{B},\mu)$ the largest extension of λ to a measure regular w.r.t. $\mathcal{K}$. Then $(X, \mathcal{B},\mu)$ is an extension of $(X, \mathcal{B}_o,\mu_o)$.

Proof. The first part has been proved above. Assume now that $\mathcal{K}_o$ is a $(\emptyset,\cup f,\cap c)$-paving. We denote by μ_{o*} and μ_* the inner measures corresponding to λ_o and λ, respectively.

(vii)
$$\mu_{o*} \le \mu_*$$

Proof. Let A be a subset of X. We have

$$\mu_{o*}(A) = \sup_{K_o \subseteq A} \lambda_o(K_o) = \sup_{K_o \subseteq A} \lambda(K_o)$$
$$\le \sup_{K \subseteq A} \lambda(K) = \mu_*(A).$$

▯

(viii)
$$\forall_{K_1 \subseteq K_2}\ \forall_{E \subseteq X}\ \mu_*(K_2 \cap E) \le \mu_*(K_1 \cap E) + \mu(K_2 \setminus K_1).$$

Proof.
$$\mu_*(K_2 \cap E) = \sup_{K \subseteq K_2 \cap E} \mu K$$
$$= \sup_{K \subseteq K_2 \cap E} \left(\mu(K \cap K_1) + \mu(K \setminus K_1)\right)$$

$$\leq \mu_*(K_1 \cap E) + \mu(K_2 \setminus K_1).$$

▯

(ix) $$A \in \mathcal{B}_o \Rightarrow A \in \mathcal{B}.$$

Proof. Consider a $K \in \mathcal{K}$. We have by (viii):

$$\mu_*(K \cap A) + \mu_*(K \setminus A)$$

$$\geq \sup_{K_o \supseteq K} \left[\left(\mu_*(K_o \cap A) - \mu(K_o \setminus K)\right) + \left(\mu_*(K_o \setminus A) - \mu(K_o \setminus K)\right) \right]$$

$$\geq \sup_{K_o \supseteq K} \left[\mu_{o*}(K_o \cap A) + \mu_{o*}(K_o \setminus A) - 2\mu(K_o \setminus K) \right]$$

$$= \sup_{K_o \supseteq K} \left[\lambda_o(K_o) - 2\mu(K_o \setminus K) \right]$$

$$\geq \lambda K - 2 \inf_{K_o \supseteq K} \mu(K_o \setminus K)$$

$$= \lambda K. \; ▯$$

(x) $$A \in \mathcal{B}_o \Rightarrow \mu_{o*}A = \mu_* A \quad (\text{i.e. } \mu_o A = \mu A).$$

Proof. $$\mu_* A = \sup_{K \subseteq A} \inf_{K_o \supseteq K} \lambda_o(K_o)$$

$$= \sup_{K \subseteq A} \inf_{K_o \supseteq K} \left[\mu_{o*}(K_o \cap A) + \mu_{o*}(K_o \setminus A) \right]$$

$$\leq \mu_{o*}(A) + \sup_{K \subseteq A} \inf_{K_o \supseteq K} \mu_*(K_o \setminus A)$$

$$\leq \mu_{o*}(A) + \sup_{K \subseteq A} \inf_{K_o \supseteq K} \mu(K_o \setminus K)$$

$$= \mu_{o*}(A). \; ▯$$

This completes the proof of the theorem. ▯

6. Construction of measures by approximation from outside and by approximation from inside. In this section we shall work with the set $\mathcal{M}_+(X;t)$ of tight measures on a topological space X.

Assume that $\hat{\beta}$ is a set function: $\mathcal{G}(X)\to\hat{R}_+$; if there exists a largest measure in $\mathcal{M}_+(X;t)$ dominated on $\mathcal{G}(X)$ by $\hat{\beta}$ (i.e. $\mu G \leq \hat{\beta}G\ \forall G$) then we denote this measure by $\hat{\mu}$.

Assume that $\check{\beta}$ is a set function: $\mathcal{K}(X)\to\hat{R}_+$; if there exists a smallest measure in $\mathcal{M}_+(X;t)$ dominating $\check{\beta}$ on $\mathcal{K}(X)$ (i.e. $\mu K \geq \check{\beta}K\ \forall K$) then we denote this measure by $\check{\mu}$.

Theorem 6.1. (i): Assume that $\hat{\beta}$: $\mathcal{G}(X)\to\hat{R}_+$ is monotone and satisfies the condition

$$G_1\cap G_2 = \emptyset \Rightarrow \hat{\beta}(G_1\cup G_2) \geq \hat{\beta}G_1 + \hat{\beta}G_2.$$

Then $\hat{\mu}$ exists.

(ii): Assume that $\check{\beta}$: $\mathcal{K}(X)\to\hat{R}_+$ is subadditive and assume that there exists at least one measure in $\mathcal{M}_+(X;t)$ dominating $\check{\beta}$ on $\mathcal{K}(X)$. Then $\check{\mu}$ exists.

Before the proof we remind the reader of a few facts. If μ_1 and μ_2 are in $\mathcal{M}_+(X;t)$ and if $X = \Phi_1\cup\Phi_2$ is a Hahn decomposition for $\mu_1-\mu_2$, i.e. if Φ_1 and Φ_2 are disjoint Borel sets with union X satisfying $\mu_1(A\cap\Phi_1) \geq \mu_2(A\cap\Phi_1)$ and $\mu_1(A\cap\Phi_2) \leq \mu_2(A\cap\Phi_2)$ for all $A \in \mathcal{B}(X)$, then $\mu_1\wedge\mu_2$ and $\mu_1\vee\mu_2$ are given by

$$(6.1)\qquad \mu_1\wedge\mu_2\ (A) = \mu_1(A\cap\Phi_2) + \mu_2(A\cap\Phi_1);\quad A \in \mathcal{B}(X),$$

$$(6.2)\qquad \mu_1\vee\mu_2\ (A) = \mu_1(A\cap\Phi_1) + \mu_2(A\cap\Phi_2);\quad A \in \mathcal{B}(X).$$

Furthermore, if the non-empty set $\mathcal{P} \subseteq \mathcal{M}_+(X;t)$ has a majorant then $\mu' = \sup\{\mu|\mu \in \mathcal{P}\}$ and $\mu'' = \inf\{\mu|\mu \in \mathcal{P}\}$ are given by

$$(6.3)\quad \mu'(A) = \sup\{\mu_1\vee\ldots\vee\mu_n(A)\,|\,n \in \hat{N},\ \mu_1,\ldots,\mu_n \in \mathcal{P}\};\quad A \in \mathcal{B}(X),$$

$$(6.4)\quad \mu''(A) = \inf\{\mu_1\wedge\ldots\wedge\mu_n(A)\,|\,n \in \hat{N},\ \mu_1,\ldots,\mu_n \in \mathcal{P}\};\quad A \in \mathcal{B}(X).$$

For the existence of μ' we may replace the condition that $\mathcal{P}$ have a majorant by the condition that the right hand side of (6.3) evaluated for $A = X$ is finite.

<u>Proof of Theorem</u> 6.1. (i): Denote by $\hat{\mathcal{P}}$ the set of measures in $\mathcal{M}_+(X;t)$ dominated on $\mathcal{G}(X)$ by $\hat{\beta}$. $\hat{\mathcal{P}}$ is non-empty since $0 \in \hat{\mathcal{P}}$. Assume now that μ_1 and μ_2 are in $\hat{\mathcal{P}}$. We shall prove that $\mu_1 \vee \mu_2 \in \hat{\mathcal{P}}$. Let $X = \Phi_1 \cup \Phi_2$ be a Hahn decomposition for $\mu_1 - \mu_2$. Consider an open set G. To $\varepsilon > 0$ we can find $K_1 \subseteq G \cap \Phi_1$ and $K_2 \subseteq G \cap \Phi_2$ such that

$$\mu_1 \vee \mu_2 \ (G) \leq \mu_1(K_1) + \mu_2(K_2) + \varepsilon .$$

Since K_1 and K_2 are disjoint compact sets contained in G, we can find disjoint open neighborhoods $N(K_1)$ and $N(K_2)$ such that $N(K_1) \cup N(K_2) \subseteq G$.

We then have

$$\begin{aligned} \mu_1 \vee \mu_2 (G) &\leq \mu_1 (N(K_1)) + \mu_2 (N(K_2)) + \varepsilon \\ &\leq \hat{\beta} (N(K_1)) + \hat{\beta} (N(K_2)) + \varepsilon \\ &\leq \hat{\beta} (N(K_1) \cup N(K_2)) + \varepsilon \\ &\leq \hat{\beta} G + \varepsilon \end{aligned}$$

Since ε was arbitrary positive and G arbitrary open, we see that $\mu_1 \vee \mu_2 \in \hat{\mathcal{P}}$. Put $\hat{\mu} = \sup\{\mu | \mu \in \hat{\mathcal{P}}\}$. Then

$$\hat{\mu}(A) = \sup\{\mu(A) \mid \mu \in \hat{\mathcal{P}}\}; \quad A \in \mathcal{B}(X),$$

from which it follows that $\hat{\mu}$ is a well-defined measure in $\mathcal{M}_+(X;t)$ and that, indeed, $\hat{\mu} \in \hat{\mathcal{P}}$.

(ii): Denote by $\check{\mathcal{P}}$ the set of measures in $\mathcal{M}_+(X;t)$ dominating $\check{\beta}$ on $\mathcal{K}(X)$. By assumption, $\check{\mathcal{P}}$ is non-empty. Let μ_1 and μ_2 be measures in $\check{\mathcal{P}}$. Consider a compact set K and a positive ε. Determine open neighborhoods $N(K \cap \Phi_2)$ and $N(K \cap \Phi_1)$ such that

$$\mu_1 \wedge \mu_2 \ (K) \geq \mu_1\big(N(K \cap \Phi_2)\big) + \mu_2\big(N(K \cap \Phi_1)\big) - \varepsilon .$$

Φ_1 and Φ_2 have the same meaning as above. By a well-known lemma, we can determine K_1 and K_2 such that $K = K_1 \cup K_2$, $K_1 \subseteq N(K \cap \Phi_2)$ and $K_2 \subseteq N(K \cap \Phi_1)$.

We then have:

$$\mu_1 \wedge \mu_2\ (K) \geq \mu_1(K_1) + \mu_2(K_2) - \varepsilon$$
$$\geq \check{\beta} K_1 + \check{\beta} K_2 - \varepsilon$$
$$\geq \check{\beta}\ (K_1 \cup K_2) - \varepsilon$$
$$= \check{\beta} K - \varepsilon.$$

In this way we see that $\mu_1 \wedge \mu_2 \in \check{\mathfrak{F}}$. It is now easy to complete the proof. ▯

Theorem 6.2. (i) (Construction of measures by approximation from outside): If $\hat{\beta}$: $\mathcal{G}(X) \to \hat{R}_+$ is monotone, additive and subadditive (i.e. $\hat{\beta}$ is a content on $\mathcal{G}(X)$), then $\hat{\mu}$ exists, and $\hat{\mu}$ is given by the formula

(6.5) $$\hat{\mu}(A) = \sup_{K \subseteq A} \inf_{G \supseteq K} \hat{\beta} G; \quad A \in \mathcal{B}(X).$$

(ii) (Construction of measures by approximation from inside): If $\check{\beta}$: $\mathcal{K}(X) \to \hat{R}_+$ is monotone, additive, subadditive (i.e. $\check{\beta}$ is a content on $\mathcal{K}(X)$), and if $\sup\{\check{\beta}K \mid K \in \mathcal{K}(X)\}$ is finite, then $\check{\mu}$ exists, and $\check{\mu}$ is given by the formula

(6.6) $$\check{\mu}(A) = \inf_{G \supseteq A} \sup_{K \subseteq G} \check{\beta} K; \quad A \in \mathcal{B}(X).$$

In (i) it is the measures of the compact sets we compute by approximation from outside. If we apply (6.5) for all subsets of X then we obtain the formula for $\hat{\mu}_*$. "Dual" remarks applies to (ii).

Proof. (i) is a special case of Theorem 2, [26].

(ii): With the given set-function $\check{\beta}$ we associate a set-function $\hat{\beta}$: $\mathcal{G}(X) \to \hat{R}_+$ defined by

$$\hat{\beta} G = \sup_{K \subseteq G} \check{\beta} K; \quad G \in \mathcal{G}(X).$$

It is not difficult to see that $\hat{\beta}$ is finite (as indicated), monotone, additive and subadditive. According to (i), the formula

$$\mu A = \sup_{K \subseteq A} \inf_{G \supseteq K} \hat{\beta} G; \quad A \in \mathcal{B}(X)$$

defines a measure in $\mathcal{M}_+(X;t)$. We shall prove that, in fact,

$$\mu A = \inf_{G \supseteq A} \beta G; \quad A \in \mathcal{B}(X).$$

When this is proved, the assertion in (ii) follows readily. We need only prove that $\mu G = \beta G$; $G \in \mathcal{G}(X)$. Let G_o be open and ε positive. Choose $K_o \subseteq G_o$ such that $\beta G_o \leq \check{\beta} K_o + \varepsilon$. Then

$$\begin{aligned} \mu G_o &\geq \inf_{G \supseteq K_o} \beta G \\ &= \inf_{G \supseteq K_o} \sup_{K \subseteq G} \check{\beta} K \\ &\geq \check{\beta} K_o \\ &\geq \beta G_o - \varepsilon, \end{aligned}$$

and it follows that $\mu G_o \geq \beta G_o$. Since the reverse inequality is obvious, we find that $\mu G_o = \beta G_o$. ▯

Note that part (ii) of the theorem just proved is an immediate consequence of Theorem 2.2 in case $\check{\beta}$ is tight.

Assume now that (μ_α) is a net on $\mathcal{M}_+(X)$ with $\limsup \mu_\alpha X < \infty$. The set-function

$$\beta G = \liminf \mu_\alpha G; \quad G \in \mathcal{G}(X)$$

satisfies the requirements of (i), Theorem 6.1, and thus β determines a measure which we shall denote by $(\hat{\mu_\alpha})$. This measure is the largest measure in $\mathcal{M}_+(X;t)$ dominated on the open sets by $\liminf \mu_\alpha(\cdot)$.

Similarly, we define a measure $(\check{\mu_\alpha})$ by considering the set function

$$\check{\beta} K = \limsup \mu_\alpha K; \quad K \in \mathcal{K}(X).$$

In case there exists no measure in $\mathcal{M}_+(X;t)$ dominating $\check{\beta}$ on $\mathcal{K}(X)$, we write $(\check{\mu_\alpha}) = \infty$.

<u>Theorem</u> 6.3. <u>In this theorem, all nets</u> (μ_α) <u>are nets on</u> $\mathcal{M}_+(X)$ <u>with</u> $\limsup \mu_\alpha X < \infty$.

(i): <u>If</u> (μ_β) <u>is a subnet of</u> (μ_α) <u>then</u> $(\hat{\mu}_\beta) \geq (\hat{\mu}_\alpha)$ <u>and</u> $(\check{\mu}_\beta) \leq (\check{\mu}_\alpha)$.

(ii): <u>If</u> (μ_α) <u>is an universal net then</u>

$$(\check{\mu}_\alpha)(A) = \inf_{G \supseteq A} \sup_{K \subseteq G} \lim \mu_\alpha K; \quad A \in \mathcal{B}(X),$$

$$(\hat{\mu}_\alpha)(A) = \sup_{K \subseteq A} \inf_{G \supseteq K} \lim \mu_\alpha G; \quad A \in \mathcal{B}(X).$$

<u>In particular, it follows that</u> $(\check{\mu}_\alpha) \leq (\hat{\mu}_\alpha)$.

(iii): <u>A necessary and sufficient condition that</u> $(\check{\mu}_\alpha) = (\hat{\mu}_\alpha)$ <u>holds for every universal net</u> (μ_α), <u>is that</u> X <u>be locally compact</u>.

<u>Proof</u>. Only the necessity part of (iii) requires proof. Assume that X is not locally compact. Then there exists $x_o \in X$ such that

$$x_o \in \bigcap \{ \overline{\complement K} \mid K \in \mathcal{K}(X) \},$$

and it follows that the filterbase $\{ \complement K \mid K \in \mathcal{K}(X) \}$ can be refined to a filter converging to x_o. When translating this fact into the language of nets, we see that there exists a net (x_α) on X such that $x_\alpha \to x_o$ and such that, for every $K \in \mathcal{K}(X)$, x_α is eventually in the completement of K. We may as well assume that (x_α) is an universal net. Now denote by μ_α a unit mass at the point x_α. It is easy to see that $(\check{\mu}_\alpha)$ is the zero measure and that $(\hat{\mu}_\alpha)$ is a unit mass at the point x_o. ▯

7. <u>On the possibility of providing a space of measures with a vague topology.</u> The space of measures we have in mind is the space $\mathcal{S}_+(X;t)$ of all measures μ in $\mathcal{M}_+(X;t)$ with $\mu X \leq 1$ where X is a topological space. It is well known that in case X is locally compact, $\mathcal{S}_+(X;t)$ can be provided with a vague topology, a prominent feature of which is that it makes $\mathcal{S}_+(X;t)$ compact. We begin by proving that it is only in the locally compact case that this can be achieved. By $\mathcal{E}$ we denote the collection of one-point measures ε_x; $x \in X$.

<u>Theorem</u> 7.1. <u>Let X be a topological space</u> (<u>Hausdorff as always</u>)

and assume that $\mathcal{P}_+(X;t)$ <u>is provided with a topology τ satisfying the following four conditions:</u>

(i): <u>The natural embedding</u> i: $X \to \mathcal{E}$ <u>is a homeomorphism,</u>

(ii): $\mathcal{P}_+(X;t)$ <u>is a Hausdorff space,</u>

(iii): $\mathcal{P}_+(X;t)$ <u>is compact,</u>

(iv): <u>The mapping</u> $\mu \to \mu(G)$ <u>of</u> $\mathcal{P}_+(X;t)$ <u>into</u> $[0,1]$ <u>is lower semi-continuous for every open subset</u> G <u>of</u> X.

<u>Then</u> X <u>must be locally compact.</u>

<u>Proof</u>. The result follows from the observation that $\mathcal{E} \cup \{0\}$ is a closed subset of $\mathcal{P}_+(X;t)$. To see this, assume that

$$\varepsilon_{x_\alpha} \overset{\tau}{\to} \mu$$

and that $\mu \neq 0$. Choose $x \in \operatorname{supp}(\mu)$. For any open neighbourhood $N(x)$ of x we have

$$0 < \mu(N(x)) \leq \liminf \varepsilon_{x_\alpha}(N(x)).$$

We conclude that $x_\alpha \in N(x)$, eventually. Thus $x_\alpha \to x$. By (i) and (ii) it follows that $\mu = \varepsilon_x$. ▯

We shall now see that conditions (i)-(iv) of Theorem 7.1 are satisfied for a notion of convergence on $\mathcal{P}_+(X;t)$. Since all the conditions (i)-(iv) can, in a natural way, be expressed in the language of nets, it is clear what we mean by this.

Let $\mu \in \mathcal{P}_+(X;t)$ and let (μ_α) be a net on $\mathcal{P}_+(X;t)$; we define $\mu_\alpha \to_t \mu$ to mean that for every universal subnet (μ_β) of (μ_α) we have

$$\widehat{(\mu_\beta)} = \mu.$$

We leave the verification of the following theorem to the reader.

<u>Theorem</u> 7.2. <u>For any topological space</u> X, <u>the notion</u> "$\to_t$" <u>defined above is a notion of convergence on</u> $\mathcal{P}_+(X;t)$ <u>satisfying</u> (i)-(iv) <u>of Theorem</u> 7.1.

<u>Lemma</u> 7.3. <u>Let</u> (μ_α) <u>be a net on</u> $\mathcal{P}_+(X;t)$. <u>If, for some</u> $\mu \in \mathcal{P}_+(X;t)$,

$$\mu \leq (\hat{\mu}_\alpha)$$

and

$$\mu X \geq \sup_K \inf_{G \supseteq K} \limsup \mu_\alpha G$$

holds, then $\mu_\alpha \to_t \mu$.

Proof. Let $(\mu_\beta)_{\beta \in E}$ be an universal subnet of $(\mu_\alpha)_{\alpha \in D}$. Then

$$\mu \leq (\hat{\mu}_\alpha) \leq (\hat{\mu}_\beta),$$

and in order to show that $\mu = (\hat{\mu}_\beta)$ it therefore suffices to prove that $\mu X \geq (\hat{\mu}_\beta)(X)$ holds. However, this follows from the considerations:

$$(\hat{\mu}_\beta)(X) = \sup_K \inf_{G \supseteq K} (\hat{\mu}_\beta)(G)$$

$$\leq \sup_K \inf_{G \supseteq K} \liminf_E \mu_\beta G$$

$$\leq \sup_K \inf_{G \supseteq K} \limsup_D \mu_\alpha G$$

$$\leq \mu X.$$

▯

The notion of convergence defined on $\mathcal{P}_+(X;t)$ can be characterized in more concrete terms:

Theorem 7.4. *Let (μ_α) be a net on $\mathcal{P}_+(X;t)$. Then (μ_α) converges in the notion $\to_t$ if and only if the equation*

(7.1) $$\inf_{G \supseteq K} \liminf \mu_\alpha G = \inf_{G \supseteq K} \limsup \mu_\alpha G$$

holds for every compact K. And when this condition is fulfilled we have $\mu_\alpha \to_t \mu$ where $\mu \in \mathcal{P}_+(X;t)$ is given by

(7.2) $$\mu A = \sup_{K \subseteq A} \inf_{G \supseteq K} \liminf \mu_\alpha G; \ A \in \mathcal{B}(X).$$

Proof. If (7.1) holds then

$$K \to \inf_{G \supseteq K} \liminf \mu_\alpha G$$

defines a semi-regular content, hence also a tight content (cf. Lemma 2.4, (ii)). According to Theorem 2.2, μ defined by (7.2) is a measure in $\mathcal{P}_+(X;t)$. It is easy to check that for any universal subnet (μ_β) of (μ_α) we have $\widehat{(\mu_\beta)} = \mu$. Thus $\mu_\alpha \overset{*}{\to} \mu$.

Assume now that $(\mu_\alpha)_{\alpha\in D}$ converges in the notion $\overset{*}{\to}$, say $\mu_\alpha \overset{*}{\to} \mu$. Clearly, the inequality

$$\mu K \leq \inf_{G\supseteq K} \liminf \mu_\alpha G$$

holds for every compact K. Assume, for the purpose of an indirect proof, that, for some $K \in \mathcal{K}(X)$, we have

$$\inf_{G\supseteq K} \limsup \mu_\alpha G > \mu K.$$

Then there exists $\varepsilon > 0$ such that, for any $G \supseteq K$, $\limsup \mu_\alpha G > \mu K+\varepsilon$ holds. Consider the set E consisting of pairs (G,α) such that $G \supseteq K$, $\alpha \in D$ and $\mu_\alpha G > \mu K+\varepsilon$. Direct E in the natural way $\big((G_1,\alpha_1) \geq (G_2,\alpha_2) \Leftrightarrow G_1 \subseteq G_2,$ $\alpha_1 \geq \alpha_2\big)$. If $\beta = (G,\alpha)$ is an element of E, we denote by μ_β the measure μ_α. Then $(\mu_\beta)_{\beta\in E}$ is a subnet of $(\mu_\alpha)_{\alpha\in D}$. Since $\mu_\beta \overset{*}{\to} \mu$, we have

$$\mu K \geq \inf_{G\supseteq K} \liminf_{\beta} \mu_\beta G,$$

but, according to our construction of the net $(\mu_\beta)_{\beta\in E}$, we also have

$$\inf_{G\supseteq K} \liminf_{\beta} \mu_\beta G \geq \mu K+\varepsilon.$$

This is a contradiction. $\square$

The last thing we shall do in this section is to define a topology on $\mathcal{P}_+(X;t)$ that almost satisfies (i)-(iv), Theorem 7.1. The topology, which we shall indicate by the letter "ν" is defined as the weakest topology on $\mathcal{P}_+(X;t)$ such that, firstly, every map $\mu \mapsto \mu G$ where $G \in \mathcal{G}(X)$ is lower semi-continuous and such that, secondly, every map $\mu \mapsto \mu K$ where $K \in \mathcal{K}(X)$ is upper semi-continuous. In other terms, we have $\mu_\alpha \overset{\nu}{\to} \mu$ if and only if

$$\mu G \leq \liminf \mu_\alpha G;\ G \in \mathcal{G}(X)$$

and

$$\mu K \geq \limsup \mu_\alpha K;\ K \in \mathcal{K}(X)$$

hold. With previously introduced notations we can express this in the form

$$\mu_\alpha \xrightarrow{\vartheta} \mu \iff (\check{\mu_\alpha}) \leq \mu \leq (\hat{\mu_\alpha}).$$

We know by Theorem 7.1 that the topology ϑ can not satisfy all of (i)-(iv), Theorem 7.1 unless X is locally compact. Theorem 6.3, or rather the proof of that theorem, explains in more explicit terms why this is so.

Before presenting the results on the topology ϑ, we find it convenient to introduce a certain class of spaces. If X is a topological space and A a subset of X, then the system $\hat{N}(A)$ of neighborhoods of A has a countable base if we can find a decreasing sequence $G_1 \supseteq G_2 \supseteq \cdots$ of open sets all containing A such that, for any open set G containing A, the inclusion $G_k \subseteq G$ holds for some k. In lack of a better name, we shall say that X <u>has property</u> (*) provided we, for any compact set K, can find a compact set K_o with $K_o \supseteq K$ such that the neighborhoodsystem $\hat{N}(K_o)$ has a countable base. Note that the class of spaces with property (*) contains all metrizable spaces, all spaces satisfying the second axiom of countability, and all locally compact spaces. The next two lemmas will make it clear how we will utilize property (*).

<u>Lemma</u> 7.5. <u>Let</u> K_o <u>be a compact subset of the topological space</u> X, <u>assume that</u> $\hat{N}(K_o)$ <u>has a countable base and let</u> $G_1 \supseteq G_2 \supseteq \cdots$ <u>be one such base. If, for each</u> $k \geq 1$, K_k <u>is a compact set with</u> $K_o \subseteq K_k \subseteq G_k$, <u>then the set</u>

$$A = \bigcup_1^\infty K_k$$

<u>is a compact set</u>.

<u>Proof</u>. Let (x_α) be an universal net on A. We shall prove that x_α converges to some point of A. If x_α converges to some point of K_o, we

are through. If not, we have ($N(\cdot)$ denoting neighbourhoods)

$$\forall_{x\in K_o} \ \exists_{N(x)} \ x_\alpha \in \complement N(x), \text{ frequently.}$$

Since (x_α) is universal we even have

$$x_\alpha \in \complement N(x), \text{ eventually.}$$

By compactness of K_o we can then find a neighbourhood $N(K_o)$ of K_o such that

$$x_\alpha \in \complement N(K_o), \text{ eventually.}$$

It follows that, for some $k \geq 1$, we have

$$x_\alpha \in \complement G_k, \text{ eventually.}$$

Since each x_α belongs to the set $A = \bigcup_1^\infty K_k$ we conclude that

$$x_\alpha \in \bigcup_{\nu=1}^{k-1} K_\nu, \text{ eventually.}$$

Clearly, there then exists $x \in \bigcup_1^{k-1} K_\nu$ such that $x_\alpha \to x$. ▯

In the next lemma we prove two results, of which only the last will be used in this section.

Lemma 7.7. Let K_o be a compact subset of X such that $\mathring{N}(K_o)$ has a countable base, and let $(\mu_n)_{n\geq 1}$ be any sequence on $\mathcal{M}_+(X;t)$. Then the following two equations are valid:

(i) $$\inf_{G\supseteq K_o} \liminf_{n\to\infty} \mu_n G = \inf_{G\supseteq K_o} \sup_{K\subseteq G} \liminf_{n\to\infty} \mu_n K,$$

(ii) $$\inf_{G\supseteq K_o} \limsup_{n\to\infty} \mu_n G = \inf_{G\supseteq K_o} \sup_{K\subseteq G} \limsup_{n\to\infty} \mu_n K.$$

Proof. (i): Let a denote the left hand side of (i). We are to prove that

$$a \leq \inf_{G\supseteq K_o} \sup_{K\subseteq G} \liminf \mu_n K$$

holds (the reverse inequality beeing trivial). In other terms, we are to prove the following:

$$\forall_{G\supseteq K_o}\ \forall_{\varepsilon>0}\ \exists_{K\subseteq G}\ \liminf \mu_n K \geq a-\varepsilon .$$

Therefore, let $G\supseteq K_o$ and $\varepsilon>0$ be given. Let

$$G_1 \supseteq G_2 \supseteq \ldots \ (\supseteq K_o)$$

denote a countable base for $\hat{N}(K_o)$ with $G_1 \subseteq G$. Since we have:

$$\forall_{k\geq 1}\ \liminf_{n\to\infty} \mu_n(G_k) \geq a,$$

we can find $n_1 < n_2 < \ldots$ such that

$$\forall_{k\geq 1}\ \forall_{n\geq n_k}\ \mu_n(G_k) > a-\varepsilon$$

holds. For each $k \geq 1$ we choose a compact set K_k with $K_o \subseteq K_k \subseteq G_k$ such that the following is true:

$$\mu_n(K_k) \geq a-\varepsilon \text{ for } n_k \leq n < n_{k+1}.$$

Put $K = \bigcup_1^\infty K_k$. K is compact and contained in G. Furthermore,

$$\liminf \mu_n K \geq a-\varepsilon$$

holds, since for $n \geq n_1$ we have $\mu_n K \geq a-\varepsilon$. This proves (i).

To prove (ii), denote the left hand side of (ii) by b and observe that what we have to prove is the following:

$$\forall_{G\supseteq K_o}\ \forall_{\varepsilon>0}\ \exists_{K\subseteq G}\ \limsup \mu_n K \geq b-\varepsilon .$$

Let $G \supseteq K_o$ and $\varepsilon > 0$ be given. We shall use the same G_k's as above. Since

$$\forall_{k\geq 1}\ \limsup_{n\to\infty} \mu_n(G_k) \geq b,$$

we can find $n_1 < n_2 < \ldots$ such that

$$\forall_{k\geq 1}\ \mu_{n_k}(G_k) > b-\varepsilon$$

holds. For each $k \geq 1$ we choose K_k with $K_o \subseteq K_k \subseteq G_k$ such that

$$\mu_{n_k}(K_k) \geq b-\varepsilon$$

holds. Put $K = \bigcup_1^\infty K_k$. K is compact and contained in G. Furthermore,

$$\limsup \mu_n K \geq b-\varepsilon$$

holds. This proves (ii). ▯

Theorem 7.7. The topology ϑ is weaker than the notion $\rightarrow_t$ in the sense that $\mu_\alpha \rightarrow_t \mu$ implies $\mu_\alpha \overset{\vartheta}{\rightarrow} \mu$. The topology ϑ satisfies conditions (i), (iii) and (iv) of Theorem 7.1. Furthermore, in case X has property (*), ϑ and $\rightarrow_t$ have the same convergent sequences; in particular, it follows that $\mathcal{P}_+(X;t)$ is "sequentially Hausdorff" in the topology ϑ (i.e. sequences have unique limits).

Proof. Only the last part of the theorem needs proof. Assume then that X has property (*) and that $\mu_n \overset{\vartheta}{\rightarrow} \mu$. By property (*) and by Lemma 7.6, (ii) we have (denoting by K_o compact sets for which $\hat{N}(K_o)$ has a countable base):

$$\mu X = \sup_{K_o} \mu K_o = \sup_{K_o} \inf_{G\supseteq K_o} \sup_{K\subseteq G} \mu K$$

$$\geq \sup_{K_o} \inf_{G\supseteq K_o} \sup_{K\subseteq G} \limsup \mu_n K$$

$$= \sup_{K_o} \inf_{G\supseteq K_o} \limsup \mu_n G$$

$$= \sup_{K} \inf_{G\supseteq K} \limsup \mu_n G.$$

Hence, by Lemma 7.3, $\mu_n \rightarrow_t \mu$ holds. ▯

Since, in nice spaces, we only know that sequences of measures are wellbehaved, it is perhaps more informative to know that $\mathcal{P}_+(X;t)$ is sequentially compact than to know that $\mathcal{P}_+(X;t)$ is compact. Therefore, we prove the following result.

Theorem 7.8. Assume that X satisfies the second axiom of countability. Then $\mathcal{P}_+(X;t)$ is sequentially compact in the topology ϑ.

__Proof__. Let $\mathcal{G} \subseteq \mathcal{G}(X)$ be a countable base closed under finite unions. Essentially, what we are going to prove is that if (μ_n) is a sequence on $\mathcal{G}_+(X;t)$ such that $\lim \mu_n G$ exists for all $G \in \mathcal{G}$, then (μ_n) converges. Let $K \in \mathcal{K}(X)$. Denote by $\mathcal{G}'$ the class of sets in $\mathcal{G}$ containing K. Then $\mathcal{G}'$ is a base for $\hat{N}(K)$. Denoting sets in $\mathcal{G}'$ by G' we have

$$\begin{aligned}\inf_{G \supseteq K} \liminf \mu_n G &= \inf_{G' \supseteq K} \liminf \mu_n G' \\ &= \inf_{G' \supseteq K} \limsup \mu_n G' \\ &= \inf_{G \supseteq K} \limsup \mu_n G,\end{aligned}$$

hence, by Theorem 7.4, (μ_n) converges in the notion $\to_t$. Then (μ_n) also converges in the topology ϑ. ▯

P A R T II

8. <u>Definition and basic properties of the weak topology.</u> Let X be a topological space and define the <u>weak topology</u> on $\mathcal{M}_+(X)$ as the weakest topology on $\mathcal{M}_+(X)$ for which every map $\mu \to \mu(f)$, where $f: X \to \hat{R}$ is bounded and upper semi-continuous, is upper semi-continuous. Convergence in the weak topology is indicated by the letter "w".

The induced topology on $\mathcal{M}_+(X;\tau)$ or $\mathcal{M}_+(X;t)$ will be called the weak topology, too. It is my feeling that the objects best suited for the study of weak convergence are the space $\mathcal{M}_+(X;t)$ for an arbitrary space X, and the space $\mathcal{M}_+(X;\tau)$ for a regular space X.

<u>Theorem</u> 8.1. (<u>Portmanteau theorem</u>). <u>Let</u> μ <u>be a measure in</u> $\mathcal{M}_+(X)$ <u>and</u> (μ_α) <u>a net on</u> $\mathcal{M}_+(X)$. <u>Consider the following eight conditions</u>:

(i): $\mu_\alpha \xrightarrow{w} \mu$;

(ii): $\limsup \mu_\alpha(f) \leq \mu(f)\ \forall f$ <u>bounded, u.s.c.</u>;

(iii): $\liminf \mu_\alpha(f) \geq \mu(f)\ \forall f$ <u>bounded, l.s.c.</u>;

(iv): $\limsup \mu_\alpha F \leq \mu F\ \forall F \in \mathcal{F}(X)$, $\lim \mu_\alpha X = \mu X$;

(v): $\liminf \mu_\alpha G \geq \mu G\ \forall G \in \mathcal{G}(X)$, $\lim \mu_\alpha X = \mu X$;

(vi): $\lim \mu_\alpha A = \mu A\ \forall$ μ-<u>continuity set</u> A;

(vii): $\lim \mu_\alpha(f) = \mu(f)\ \forall$ <u>bounded</u> μ-<u>continuity function</u> f;

(viii): $\lim \mu_\alpha(f) = \mu(f)\ \forall$ <u>bounded continuous</u> f.

<u>Then the first five conditions are all equivalent and each of them implies the remaining three conditions</u>.

<u>If X is completely regular and if</u> $\mu \in \mathcal{M}_+(X;\tau)$ <u>then all eight</u>

<u>conditions given above are equivalent and, furthermore, if we consider a fixed uniform structure</u> $\mathcal{U}$ <u>on</u> X <u>then we can add the following condition equivalent with those above:</u>

(ix): $\lim \mu_\alpha(f) = \mu(f)\ \forall$ <u>bounded</u> f, <u>uniformly continuous w.r.t.</u> $\mathcal{U}$.

<u>Proof</u>. (i) and (ii) are equivalent by definition. Clearly, (ii) and (iii) are equivalent.

We shall now prove that (ii) and (iii) imply (vii). If f is bounded and μ-continuous then, with the notation of P9 , $\mu(f_*) = \mu(f^*)$ and, by (ii) and (iii) we have

$$\mu(f_*) \leq \liminf \mu_\alpha(f_*) \leq \limsup \mu_\alpha(f^*) \leq \mu(f^*).$$

Clearly, (vii) implies (vi) as well as (viii).

If (ii) holds then the first condition in (iv) holds and, since (ii) implies (viii), we also find that the second condition in (iv) holds.

Clearly, (iv) and (v) are equivalent.

We shall now prove that (iv) implies (ii). First assume that $0 < f < 1$ and that f is u.s.c. Then we have, for every $k \in \hat{N}$ (see P15):

$$\begin{aligned}\limsup \mu_\alpha(f) &\leq \limsup\left\{\mu_\alpha X/k + 1/k\ \Sigma_1^k\ \mu_\alpha(\{f \geq \nu/k\})\right\}\\ &\leq \limsup \mu_\alpha X/k + 1/k\ \Sigma_1^k \limsup \mu_\alpha(\{f \geq \nu/k\})\\ &\leq \mu X/k + 1/k\ \Sigma_1^k\ \mu(\{f \geq \nu/k\})\\ &\leq \mu X/k + \mu(f),\end{aligned}$$

and it follows that $\limsup \mu_\alpha(f) \leq \mu(f)$. If $a < f < b$ and if f is u.s.c. then, by what we have just proved,

$$\limsup \mu_\alpha\Big((f-a)/(b-a)\Big) \leq \mu\Big((f-a)/(b-a)\Big),$$

from which it follows that

$$\limsup\big(\mu_\alpha(f) - a\ \mu_\alpha(X)\big) \leq \mu(f) - a\ \mu(X).$$

Since $\lim \mu_\alpha X = \mu X$ we conclude that $\limsup \mu_\alpha(f) \leq \mu(f)$.

For the rest of the proof we assume that X is completely regular and that $\mu \in \mathcal{M}_+(X;\tau)$.

Assume that (ix) holds. We shall prove that then (iv) holds. $\lim \mu_\alpha X = \mu X$ obviously holds. Let $F \in \mathcal{F}(X)$ and consider the family $\mathcal{F}$ of functions f uniformly continuous w.r.t. $\mathcal{U}$ and satisfying the inequalities $0 \leq f \leq 1$ and, furthermore, the requirement that $f(x) = 1$ for all $x \in F$. Then $\mathcal{F}$ is filtering to the left and, according to a result of Weil (cf. [28], p.14), we have

$$\inf\{f \mid f \in \mathcal{F}\} = 1_F.$$

We now have

$$\mu F = \inf_{f\in\mathcal{F}} \mu(f) = \inf_{f\in\mathcal{F}} \lim \mu_\alpha(f)$$

$$\geq \limsup \mu_\alpha F.$$

Having seen that (ix) implies (iv), it is clear that (vii) as well as (viii) imply (iv).

The proof that (vi) implies (iv) utilizes the same idea as the one in the proof of (ix) $\Rightarrow$ (iv) and is based on P16. ▯

9. <u>Compactness in the weak topology.</u> A net (μ_α) on $\mathcal{M}_+(X)$ is said to be <u>tight</u> if the condition

$$\inf_{K\in\mathcal{K}(X)} \limsup_\alpha \mu_\alpha(\complement K) = 0$$

is satisfied. And (μ_α) is said to be τ-<u>smooth</u> if the equation

$$\inf_{F\in\mathcal{F}} \limsup_\alpha \mu_\alpha F = 0$$

holds for every subclass $\mathcal{F}$ of $\mathcal{F}(X)$ filtering downwards to the empty set ($\mathcal{F} \downarrow \emptyset$).

A subset $\mathcal{P}$ of $\mathcal{M}_+(X)$ is <u>tight</u> if

$$\inf_{K\in\mathcal{K}(X)} \sup_{\mu\in\mathcal{P}} \mu(\complement K) = 0.$$

And $\mathcal{P}$ is τ-<u>smooth</u> if

$$\inf_{F\in\mathcal{F}} \sup_{\mu\in\mathcal{P}} \mu F = 0$$

for every $\mathcal{F} \subseteq \mathcal{F}(X)$ with $\mathcal{F} \downarrow \emptyset$.

We borrow from Topsøe, [26] the following results on compactness. We collect in the first theorem the basic results on compactness in the space $\mathcal{M}_+(X;t)$ and in the second we collect the results concerned with $\mathcal{M}_+(X;\tau)$

Theorem 9.1.

(i): Let (μ_α) be a net on $\mathcal{M}_+(X;t)$ with $\limsup \mu_\alpha X < \infty$. Then (μ_α) is compact in the space $\mathcal{M}_+(X;t)$ if and only if, for every subclass $\mathcal{G}$ of $\mathcal{G}(X)$ which dominates $\mathcal{K}(X)$ (i.e. $\forall K \; \exists G \in \mathcal{G} : G \supseteq K$) we have

(9.1) $$\inf_{\mathcal{G}'} \limsup_{\alpha} \min_{G \in \mathcal{G}'} \mu_\alpha(\complement G) = 0$$

where the infimum is taken over all finite subclasses $\mathcal{G}'$ of $\mathcal{G}$.

(ii): Let again (μ_α) be a net on $\mathcal{M}_+(X;t)$ with $\limsup \mu_\alpha X < \infty$. If (μ_α) is tight then (μ_α) is compact, and if (μ_α) is compact then (μ_α) is τ-smooth.

(iii): Let $\mathcal{P}$ be a subset of $\mathcal{M}_+(X;t)$ with $\sup\{\mu X \mid \mu \in \mathcal{P}\} < \infty$. Then $\mathcal{P}$ is net-compact if and only if, for every $\mathcal{G} \subseteq \mathcal{G}(X)$ dominating $\mathcal{K}(X)$ we have

(9.2) $$\inf_{\mathcal{G}'} \sup_{\mu \in \mathcal{P}} \min_{G \in \mathcal{G}'} \mu(\complement G) = 0$$

where the infimum is over all finite subclasses $\mathcal{G}'$ of $\mathcal{G}$.

If $\mathcal{P}$ is tight then $\mathcal{P}$ is net-compact (in fact relatively compact), and if $\mathcal{P}$ is net-compact then $\mathcal{P}$ is τ-smooth.

(iv): If X is either locally compact or metrizable with a complete metric then every compact subset of $\mathcal{M}_+(X;t)$ is tight.

Theorem 9.2. Let X be a regular space and consider $\mathcal{M}_+(X;\tau)$ with the weak topology. Then a net (μ_α) on $\mathcal{M}_+(X;\tau)$ with $\limsup \mu_\alpha X < \infty$ is compact if and only if it is τ-smooth. And a subset $\mathcal{P}$ of $\mathcal{M}_+(X;\tau)$ with $\sup\{\mu X \mid \mu \in \mathcal{P}\} < \infty$ is relatively compact if and only if it is τ-smooth.

Even though the proofs of these results can be found in the above mentioned paper, we do wish to repeat the proof of one of the central statements. We choose to prove the "if-part" of Theorem 9.1, (iii). Assume then that $\mathcal{P} \subseteq \mathcal{M}_+(X;t)$ satisfies the condition (9.2) and the condition $\sup\{\mu X \mid \mu \in \mathcal{P}\} < \infty$. Let (μ_α) be an universal net on $\mathcal{P}$. We shall prove that (μ_α) converges weakly to some measure in $\mathcal{M}_+(X;t)$. Consider the set function β: $\mathcal{G}(X) \to \hat{R}_+$ defined by

$$\beta(G) = \lim \mu_\alpha(G); \; G \in \mathcal{G}(X).$$

The hypothesis of Theorem 6.2, (i) are fulfilled, hence the formula

$$\mu A = \sup_{K \subseteq A} \inf_{G \supseteq K} \beta(G); \quad A \in \mathcal{B}(X)$$

defines a measure in $\mathcal{M}_+(X;t)$. We shall prove that $\mu_\alpha \xrightarrow{w} \mu$. Since

$$\mu G \leq \beta G = \lim \mu_\alpha G; \; G \in \mathcal{G}(X),$$

it follows by Theorem 8.1, that we only have to prove that $\mu X \geq \beta X$ holds. If this were not so, we would be able to find $\varepsilon > 0$ such that to any K there would exist $G \supseteq K$ with $\beta(G) < \beta(X) - \varepsilon$. In other terms, we could find a family $(G_K)_{K \in \mathcal{K}(X)}$ such that $G_K \supseteq K$ for each K and such that

$$\lim \mu_\alpha(\complement G_K) > \varepsilon$$

for each K. Then

$$\mu_\alpha(\complement G_K) \geq \varepsilon, \text{ eventually}$$

holds for each K. It follows that, for each finite set of K's, say $K_1, \ldots, K_n$, we have

$$\min_{\nu=1,\ldots,n} \mu_\alpha(\complement G_{K_\nu}) \geq \varepsilon, \text{ eventually.}$$

This contradiction proves the assertion.

It is easy to see that those spaces X for which the compact <u>nets</u> on $\mathcal{M}_+(X;t)$ are the same as the tight nets (with $\limsup \mu_\alpha X < \infty$) are precisely the same as the locally compact spaces (employ the idea in the proof of Theorem 6.3). If we ask the corresponding question for <u>subsets</u> of $\mathcal{M}_+(X;t)$, much less is known. The best result is that of

Prohorov stating that the complete metrizable spaces are well behaved in this respect (cf. (iv) of Theorem 9.1). If we consider sets consisting of sequences we obtain the following fascinating result, essentially due to LeCam (cf. Theorem 4 of [15]).

Theorem 9.3. Assume that X satisfies condition (*) of section 7. If $(\mu_n)_{n\geq 1}$ is a sequence on $\mathcal{M}_+(X;t)$ converging weakly to a measure $\mu \in \mathcal{M}_+(X;t)$, then the set $\mathcal{P} = \{\mu_n \mid n \geq 1\}$ is tight.

Proof. To $\varepsilon > 0$ we choose $K_1 \in \mathcal{K}(X)$ such that $\mu(\complement K_1) < \varepsilon$. Denote by K_0 a compact set containing K_1 such that $\mathfrak{N}(K_0)$ has a countable base. By Lemma 7.6, (i) we have

$$\mu K_0 = \inf_{G \supseteq K_0} \mu G \leq \inf_{G \supseteq K_0} \liminf \mu_n G$$

$$= \inf_{G \supseteq K_0} \sup_{K \subseteq G} \liminf \mu_n K$$

$$\leq \sup_{K} \liminf \mu_n K.$$

Therefore, there exists $K \in \mathcal{K}(X)$ such that

$$\liminf \mu_n K \geq \mu K_0 - \varepsilon.$$

Since $\mu_n X \to \mu X$ this implies that

$$\limsup \mu_n(\complement K) \leq \mu(\complement K_0) + \varepsilon \leq 2\varepsilon,$$

and then we can of course find $K_2 \in \mathcal{K}(X)$ such that $\mu_n(\complement K_2) \leq 2\varepsilon$ holds for all $n \geq 1$. ▯

10. Criteria for weak convergence. There are several types of criteria for weak convergence $\mu_\alpha \xrightarrow{w} \mu$. In some of them, we assume that the net (μ_α) is compact, in others not. And in some criteria we do not assume that the "target-measure" μ is known. However, we have not found it possible to establish an useful criteria if we neither have a compactness assumption nor a target-measure. Some of the criteria below are

"criteria without loss" in the sense that if μ_α does converge to μ then the conditions of the criteria are satisfied.

Theorem 10.1. Let $\mathcal{A}$ be a $(\cup f, \cap f)$ paving on X separating points T_2 (see P), (μ_α) a compact net on the space $\mathcal{M}_+(X;t)$ and μ a measure in $\mathcal{M}_+(X;t)$. If

$$\lim \mu_\alpha X = \mu X$$

and

$$\limsup \mu_\alpha(\mathring{A}) \leq \mu(\bar{A}) \; \forall A \in \mathcal{A}$$

hold, then μ_α converges weakly to μ.

Proof. If (μ_β) is a weakly convergent subnet of (μ_α), say $\mu_\beta \xrightarrow{w} \mu_1$ then $\mu_1 X = \mu X$ holds and, for every $A \in \mathcal{A}$ we have

$$\mu_1 \mathring{A} \leq \liminf_\beta \mu_\beta \mathring{A} \leq \limsup_\alpha \mu_\alpha \mathring{A} \leq \mu \bar{A}.$$

By P19 it follows that $\mu_1 = \mu$. ▯

Theorem 10.2. Let $\mathcal{A}$ be a $(X, \cap f)$-paving on X consisting of Borel sets and assume that $\alpha(\mathcal{A})$, the algebra spanned by $\mathcal{A}$, separates points T_2. If (μ_α) is a compact net in the space $\mathcal{M}_+(X;t)$, and if $\lim \mu_\alpha A$ exists for every $A \in \mathcal{A}$, then μ_α converges weakly in $\mathcal{M}_+(X;t)$.

Proof. According to P11, $\lim \mu_\alpha A$ exists for every $A \in \alpha(\mathcal{A})$. Now, let (μ_β) and (μ_γ) be two convergent subnets of (μ_α), say $\mu_\beta \xrightarrow{w} \mu_1$ and $\mu_\gamma \xrightarrow{w} \mu_2$. Then, for every $A \in \alpha(\mathcal{A})$, we have

$$\mu_1 \mathring{A} \leq \liminf_\beta \mu_\beta \mathring{A} \leq \lim_\alpha \mu_\alpha A$$

$$\leq \limsup_\gamma \mu_\gamma \bar{A} \leq \mu_2 \bar{A}$$

and, since $\mu_1 X = \mu_2 X$ clearly holds, it follows by P19 that $\mu_1 = \mu_2$. ▯

We mention the following corollary without giving the simpel details of the proof.

Corollary 10.3. Assume that there exists a countable class of sub-

<u>sets of</u> X <u>separating points</u> T_2. <u>Then every net-compact subset of</u> $\mathcal{M}_+(X;t)$ <u>is sequentially compact.</u>

<u>Theorem</u> 10.4. <u>Assume that</u> $\mathcal{A} \subseteq \mathcal{B}(X)$ <u>is closed under finite unions and that</u> $\mathcal{A}$ <u>separate points and closed sets</u> T_1. <u>Let</u> (μ_α) <u>be a net on</u> $\mathcal{M}_+(X)$ <u>and</u> μ <u>a measure in</u> $\mathcal{M}_+(X;\tau)$. <u>If</u>

$$\lim \mu_\alpha X = \mu X$$

<u>and</u>

$$\liminf \mu_\alpha A \geq \mu \mathring{A} \quad \forall A \in \mathcal{A}$$

<u>hold, then</u> μ_α <u>converges weakly to</u> μ.

<u>Proof</u>. Let $G \in \mathcal{G}(X)$. Consider the class

$$\mathcal{G} = \{\mathring{A} \mid A \in \mathcal{A}, A \subseteq G\}.$$

Then $\mathcal{G} \uparrow G$, and we have

$$\mu G = \sup_{\mathring{A} \in \mathcal{G}} \mu \mathring{A} \leq \sup_{\mathring{A} \in \mathcal{G}} \liminf \mu_\alpha A \leq \liminf \mu_\alpha G.$$

We can now conclude that $\mu_\alpha \xrightarrow{w} \mu$. ∎

Employing the result of P11 we obtain the following corollary to Theorem 10.4:

<u>Corollary</u> 10.5. <u>Let</u> $\mathcal{A}$ <u>be a</u> $(X,\cap f)$<u>-paving on</u> X <u>consisting of Borel-sets and assume that</u> $\alpha(\mathcal{A})$ <u>separates points and closed sets</u> T_1. <u>If</u> (μ_α) <u>is a net on</u> $\mathcal{M}_+(X)$ <u>and</u> μ <u>a measure in</u> $\mathcal{M}_+(X;\tau)$ <u>such that</u>

$$\lim \mu_\alpha A = \mu A$$

<u>for all</u> $A \in \mathcal{A}$, <u>then</u> μ_α <u>converges weakly to</u> μ.

11. <u>On the structure of</u> $\mathcal{M}_+(X)$. Perhaps we could begin by remarking that the algebraic and topological structure of $\mathcal{M}_+(X)$ goes well together in the sense that mappings like $(\mu_1,\mu_2,a_1,a_2) \to a_1\mu_1+a_2\mu_2$ are continuous.

Our first result is concerned with the space $\mathcal{M}_+^1$ of probability measures. $i: X \to \mathcal{M}_+^1(X)$ denotes the natural imbedding of X in $\mathcal{M}_+^1(X)$. $\mathcal{E} = i(X)$ is the set of point masses ε_x; $x \in X$.

Theorem 11.1.

(i): $i: X \to \mathcal{E}$ is a homeomorphism.

(ii): $\mathcal{E}$ is a closed subset of $\mathcal{M}_+^1(X;\tau)$.

(iii): The probability measures with finite support are dense in $\mathcal{M}_+^1(X)$, in symbols

$$\overline{co}(\mathcal{E}) = \mathcal{M}_+^1(X).$$

(iv): The measures in $\mathcal{E}$ are the only 0,1-measures in $\mathcal{M}_+^1(X;\tau)$.

(v): $\mathcal{E}$ is the set of extremalpoints of $\mathcal{M}_+^1(X;r,\tau)$.

In (iii), $\overline{co}$ indicates "closed convex hull of".

Proof. (i) is trivial. The proofs of (ii) and (iv) resemble that of Theorem 7.1. To prove (iii), choose to each non-empty subset A of X a point x_A in A. Let D be the set of divisions of X in finitely many non-empty Borel-sets and direct D by "subdivision". Given $\mu \in \mathcal{M}_+^1(X)$ and $\alpha = (A_1,\dots,A_n)$ in D, define μ_α by

$$\mu_\alpha = \Sigma_1^n \mu A_i \cdot \varepsilon_{x_{A_i}}$$

For any Borel set A we have $\mu_\alpha A \to \mu A$. In particular, $\mu_\alpha \xrightarrow{w} \mu$ follows.

To prove (v) we have to prove the inclusion

$$ext\left(\mathcal{M}_+^1(X;r,\tau)\right) \subseteq \mathcal{E}$$

(the reverse inclusion is trivial). Assume that $\mu \notin \mathcal{E}$ (and that $\mu \in \mathcal{M}_+^1(X;r,\tau)$). By (iv) we can find $A \in \mathfrak{B}(X)$ with $0 < \mu A < 1$. Denote by $\mu_1[\mu_2]$ the restriction of μ to A [$\complement A$]. Then μ_1 and μ_2 are measures in $\mathcal{M}_+(X;r,\tau)$ (see P14) and the identity

$$\mu = \mu A \cdot \left((1/\mu A)\cdot\mu_1\right) + \mu(\complement A)\cdot\left(\left(1/\mu(\complement A)\right)\mu_2\right)$$

shows that μ is not an extreme point of $\mathcal{M}_+^1(X;r,\tau)$. ∎

If a topological property is inherited on closed subsets then it follows by (ii) above that X has the property in case $\mathcal{M}^1_+(X;\tau)$ or $\mathcal{M}^1_+(X;t)$ has the property.

We collect in one theorem the remaining results of this section.

Theorem 11.2.

(i): $\mathcal{M}_+(X;t)$ for X arbitrary and $\mathcal{M}_+(X;\tau)$ for X regular are Hausdorff spaces.

(ii): If X is separable then so is $\mathcal{M}_+(X)$.

(iii): $\mathcal{M}_+(X;\tau)$ is second countable if and only if X is so.

(iv): $\mathcal{M}_+(X;\tau)$ is regular if and only if X is so.

(v): $\mathcal{M}_+(X;\tau)$ is completely regular if and only if X is so.

Proof. (i): For X arbitrary, $\mathfrak{D}(X)$ separates points T_2, and for X regular, $\mathfrak{D}(X)$ separates points and closed sets T_2. Appealing to the results of P19 it is easy to see that, in each of the two cases, $\mu_\alpha \xrightarrow{w} \mu_1$ together with $\mu_\alpha \xrightarrow{w} \mu_2$ implies $\mu_1 = \mu_2$.

(ii): Let E be a countable dense subset of X and denote by $sp^*(\mathcal{E}_E)$ the set of measures

$$\Sigma_1^n\, q_i \cdot \varepsilon_{x_i}$$

where $n \geq 1$, where the q's are non-negative rationals, and where the x's are points in E. $sp^*(\mathcal{E}_E)$ is a countable dense subset of $\mathcal{M}_+(X)$.

(iii): The "only if" part is obvious. Assume now that X is second countable and denote by $\mathcal{G}^*$ a countable base for $\mathcal{G}(X)$ closed under finite unions. Elements of $\mathcal{G}^*$ will be indicated by G^*, G^*_1 etc. Note that, for any $\mu \in \mathcal{M}_+(X;\tau)$ and any $G \in \mathcal{G}(X)$, we have

(11.1) $$\mu G = \sup\{\mu G^* \mid G^* \subseteq G\}.$$

Denote by $\mathcal{K}$ the class of finite intersections of sets each one of which is either a set in $\mathcal{G}^*$ or else a set whose complement is in $\mathcal{G}^*$. Choose, for every non-empty set H in $\mathcal{K}$ a point x_H in H. Put

$$E = \{x_H \mid H \in \mathcal{K},\ H \neq \emptyset\}.$$

Then E is a countable dense subset of X. From the proof of (ii) we have that $sp^*(\mathscr{E}_E)$ is a countable dense subset of $\mathcal{M}_+(X)$.

Now, let μ be a measure in $\mathcal{M}_+(X)$, $G_1^*,\dots,G_n^*$ finitely many sets in $\mathscr{G}^*$ and ε a positive number. Consider the division $X = (H_1,\dots,H_m)$ of X induced by $G_1^*,\dots,G_n^*$. The sets H_i are non-empty elements of $\mathscr{H}$, and m is at most equal to 2^n. Choose $q_1,\dots,q_m$ non-negative rationals such that

$$|\mu H_i - q_i| < \varepsilon/m; \ i=1,\dots,m.$$

Put

$$\nu = \Sigma_1^m q_i \cdot \varepsilon_{x_{H_i}}.$$

Then $\nu \in sp^*(\mathscr{E}_E)$, $|\mu X-\nu X| < \varepsilon$ and, for each $i=1,\dots,m$, $|\mu G_i^*-\nu G_i^*| < \varepsilon$. What we have proved can shortly be expressed as follows:

(11.2) $$\underset{\mu\in \mathcal{M}_+(X)}{\forall}\ \underset{G_1^*,\dots,G_n^*}{\forall}\ \underset{\varepsilon>0}{\forall}\ \underset{\nu\in sp^*(\mathscr{E}_E)}{\exists} |\mu X-\nu X|<\varepsilon,\ \max_{1\le i\le n} |\mu G_i^*-\nu G_i^*| < \varepsilon.$$

Consider now the class $N\big(sp^*(\mathscr{E}_E),\ \mathscr{G}^*,\mathbb{Q}_+\big)$ consisting of all open subsets of $\mathcal{M}_+(X;\tau)$ of the form

$$N(\nu,G_1^*,\dots,G_n^*,\varepsilon) = \{\mu \mid \mu G_i^*>\nu G_i^*-\varepsilon; i=1,\dots,n, |\mu X-\nu X| < \varepsilon\}$$

where $\nu \in sp^*(\mathscr{E}_E)$, $G_1^*,\dots,G_n^*$ are finitely many sets in $\mathscr{G}^*$ and $\varepsilon \in \mathbb{Q}_+$, the positive rationals. Clearly, $N\big(sp^*(\mathscr{E}_E),\ \mathscr{G}^*,\mathbb{Q}_+\big)$ is countable. To prove that it is a base, consider an open subset $\mathscr{N}$ of $\mathcal{M}_+(X;\tau)$ and a measure $\mu_0 \in \mathscr{N}$. We can then find finitely many open sets $G_1,\dots,G_n$ and an $\varepsilon > 0$ such that $N(\mu_0,G_1,\dots,G_n,\varepsilon) \subseteq \mathscr{N}$ where

$$N(\mu_0,G_1,\dots,G_n,\varepsilon) = \{\mu \mid \mu G_i>\mu_0 G_i-\varepsilon;\ i=1,\dots,n, |\mu X-\mu_0 X| < \varepsilon\}$$

By (11.1), we can find $G_1^*,\dots,G_n^*$ in $\mathscr{G}^*$ and $\varepsilon^* \in \mathbb{Q}_+$ such that

$$N(\mu_0,G_1^*,\dots,G_n^*,\varepsilon^*) \subseteq N(\mu_0,G_1,\dots,G_n,\varepsilon).$$

By (11.2) we see that $\nu \in sp^*(\mathscr{E}_E)$ cen be found such that

$$\mu_0 \in N(\nu,G_1^*,\dots,G_n^*,\varepsilon^*/2) \subseteq N(\mu_0,G_1^*,\dots,G_n^*,\varepsilon^*).$$

All in all, we have found a set in $N(sp^*(\mathcal{E}_E), \mathcal{G}^*, \mathbb{Q}_+)$ containing μ_o and contained in $\mathcal{N}$. This argument proves that $N(sp^*(\mathcal{E}_E), \mathcal{G}^*, \mathbb{Q}_+)$ is a base for $\mathcal{M}_+(X;\tau)$.

(iv): The "only if" part is obvious. Assume now that X is regular. Let μ_o be a measure in $\mathcal{M}_+(X;\tau)$ and $N(\mu_o)$ a neighbourhood of μ_o. We may assume that $N(\mu_o)$ is of the form

$$N(\mu_o) = \{\mu \mid \mu G_i > \mu_o G_i - \varepsilon;\ 1 \leq i \leq n,\ |\mu X - \mu_o X| < \varepsilon\},$$

where $G_1, \ldots, G_n$ are finitely many open sets and ε is positive. Choose open sets $G_1^*, \ldots, G_n^*$ such that $\overline{G_i^*} \subseteq G_i$; $1 \leq i \leq n$ and such that $\mu_o G_i^* > \mu_o G_i - \varepsilon/2$; $1 \leq i \leq n$ (see P16). Consider the neighbourhood

$$N^*(\mu_o) = \{\mu \mid \mu G_i^* > \mu_o G_i^* - \varepsilon/2;\ 1 \leq i \leq n,\ |\mu X - \mu_o X| < \varepsilon/2\}.$$

Assume that μ is a measure in the closure of $N^*(\mu_o)$. Then there exists a net (μ_α) on $N^*(\mu_o)$ converging weakly to μ. For each $i=1,\ldots,n$ we have

$$\mu G_i \geq \mu \overline{G_i^*} \geq \limsup \mu_\alpha \overline{G_i^*} \geq \limsup \mu_\alpha G_i^*$$

$$\geq \mu_o G_i^* - \varepsilon/2 > \mu_o G_i - \varepsilon,$$

and we also have

$$|\mu X - \mu_o X| = \lim |\mu_\alpha X - \mu_o X| < \varepsilon.$$

It follows that

$$\overline{N^*(\mu_o)} \subseteq N(\mu_o).$$

(v): The "only if" part is obvious. Assume now that X is completely regular. Appealing to the equivalence (i) $\Leftrightarrow$ (viii) of Theorem 8.1., we may proceed as in the proof of Theorem II.1 in Varadarajan, [21] - or we may remark that $\mathcal{M}_+(X;\tau)$ is uniformizable, hence completely regular. ▯

12. <u>A problem related to questions of uniformity.</u> In this and the remaining sections we shall discuss certain problems on weak convergen-

ce in case the underlying space X is completely regular. Most of the results will be formulated by referring to a fixed uniform structure $\mathcal{U}$ for X, but the problems themselves are formulated without reference to any uniformity.

The main problem. Given is a measure $\mu \in \mathcal{M}_+(X)$ and a net $(\mathcal{F}_i)_{i\in I}$ where, for each $i \in I$, $\mathcal{F}_i$ is a non-empty subset of $\mathcal{B}(X;[0,1])$, the set of measurable functions $X \to [0,1]$. The problem is to compute the quantity $\xi = \xi((\mathcal{F}_i),\mu)$ defined by

$$(12.1) \qquad \xi = \sup_{(\mu_\alpha, i_\alpha)} \limsup_{\alpha} \sup_{f\in\mathcal{F}_{i_\alpha}} |\mu_\alpha(f) - \mu(f)| ,$$

where the first supremum is taken over all nets $(\mu_\alpha, i_\alpha)_{\alpha\in D}$ on $\mathcal{M}_+(X)\times I$ such that $\mu_\alpha \xrightarrow{w} \mu$ and such that $(i_\alpha)_{\alpha\in D}$ is a subnet of I.

We also define quantities ξ^+ and ξ^- by

$$(12.2) \qquad \xi^+ = \sup_{(\mu_\alpha, i_\alpha)} \limsup_{\alpha} \sup_{f\in\mathcal{F}_{i_\alpha}} \{\mu_\alpha(f) - \mu(f)\} ,$$

$$(12.3) \qquad \xi^- = \sup_{(\mu_\alpha, i_\alpha)} \limsup_{\alpha} \sup_{f\in\mathcal{F}_{i_\alpha}} \{\mu(f) - \mu_\alpha(f)\} .$$

It is elementary to check that

$$(12.4) \qquad \xi = \max(\xi^+, \xi^-) .$$

If $(\mathcal{A}_i)_{i\in I}$ is a net on the non-empty subsets of $\mathcal{B}(X)$, we define $\xi((\mathcal{A}_i),\mu)$, $\xi^+((\mathcal{A}_i),\mu)$ and $\xi^-((\mathcal{A}_i),\mu)$ in the obvious way identifying a subset of X with its indicator-function.

The fact that $(\mu_\alpha, i_\alpha)_{\alpha\in D}$ is a net on $\mathcal{M}_+(X)\times I$ with $\mu_\alpha \xrightarrow{w} \mu$ and $(i_\alpha)_{\alpha\in D}$ a subnet of I is shortly expressed by saying that (μ_α, i_α) is a subnet of $\hat{N}(\mu)\times I$ ($\hat{N}(\mu)$ denoting the neighbourhoodsystem for μ).

In particular, we shall investigate the case $\xi = 0$. If I consists of just one element, in which case we consider only one class $\mathcal{F} \subseteq \mathcal{B}(X;[0,1])$, we find that $\xi(\mathcal{F},\mu) = 0$ holds if and only if

$$\lim_{\alpha} \sup_{f \in \mathfrak{F}} |\mu_\alpha(f)-\mu(f)| = 0$$

holds for any net on $\mathcal{M}_+(X)$ converging weakly to μ. Such classes are called μ-<u>uniformity classes.</u>

We shall now introduce the notions to be used in the solution of the "$\mathfrak{F}$-problem". For uniformities we use the notation of Kelley, [13]. Elements of an uniformity $\mathcal{U}$ for X will preferable be denoted by the letters U and V . We remind the reader of the notations

$$U \circ V = \{(x,z) | (x,y) \in V , (y,z) \in U \text{ for some } y\},$$

$$U[A] = \{y | (x,y) \in U \text{ for some } x \in A\}.$$

Given $f: X \to \hat{R}$ and $A \subseteq X$ we denote by $\omega_f(A)$ the oscillation of f over A i.e.

$$\omega_f(A) = \sup\{|fx - fy| \mid x,y \in A\}.$$

Given $f: X \to \hat{R}$, $U \in \mathcal{U}$ and $\varepsilon > 0$, we define $\partial_{U,\varepsilon}(f)$, the U,ε-<u>boundary of</u> f, by

(12.5) $$\partial_{U,\varepsilon}(f) = \{x \mid \omega_f(U^{-1}[x]) > \varepsilon\}.$$

Furthermore, we define sets $\partial^+_{U,\varepsilon}(f)$ and $\partial^-_{U,\varepsilon}(f)$ by

(12.6) $$\partial^+_{U,\varepsilon}(f) = \{x \mid f(x') > f(x) + \varepsilon \text{ for some } x' \in U^{-1}[x]\},$$

(12.7) $$\partial^-_{U,\varepsilon}(f) = \{x \mid f(x') < f(x) - \varepsilon \text{ for some } x' \in U^{-1}[x]\}.$$

If $A \subseteq X$ and $U \in \mathcal{U}$ we define $\partial_U(A)$, $\partial^+_U(A)$ and $\partial^-_U(A)$ by

(12.8) $$\partial_U(A) = U[A] \cap U[\complement A],$$

(12.9) $$\partial^+_U(A) = U[A] \setminus A,$$

(12.10) $$\partial^-_U(A) = U[\complement A] \setminus \complement A.$$

Note that, for $0 < \varepsilon < 1$, $\partial_U(A) = \partial_{U,\varepsilon}(1_A)$, $\partial^+_U(A) = \partial^+_{U,\varepsilon}(1_A)$ and $\partial^-_U(A) = \partial^-_{U,\varepsilon}(1_A)$. Since $\partial^+_U A = \partial_U(A) \setminus A$ and $\partial^-_U A = \partial_U(A) \cap A$, the sets $\partial^+_U A$ and

$\partial_U^- A$ are disjoint sets with union $\partial_U A$.

We now list some properties of U,ε-boundaries all of which are simple to establish.

(12.11) $$\partial_{U,2\varepsilon} f \subseteq \partial^+_{U,\varepsilon} f \cup \partial^-_{U,\varepsilon} f \subseteq \partial_{U,\varepsilon} f .$$

(12.12) $$D(f) = \bigcup_{\varepsilon} \bigcap_{U} \partial_{U,\varepsilon} f .$$

(12.13) $$\{ f < f^* \} = \bigcup_{\varepsilon} \bigcap_{U} \partial^+_{U,\varepsilon} f .$$

(12.14) $$\{ f_* < f \} = \bigcup_{\varepsilon} \bigcap_{U} \partial^-_{U,\varepsilon} f .$$

(12.15) $$U[\partial_{V,\varepsilon} f] \subseteq \partial_{U\cdot V,\varepsilon} f .$$

(12.16) $$\overline{\partial_{U,\varepsilon} f} \subseteq \partial_{V\cdot U,\varepsilon} f \quad \forall V \in \mathcal{U}$$

(12.12) and (12.16) implies that

(12.17) $$D(f) = \bigcup_{\varepsilon} \bigcap_{U} \overline{\partial_{U,\varepsilon} f} ,$$

from which it follows, as already noticed in P , that D(f) is a Borel-set.

If U is open then $\partial_{U,\varepsilon} f$ is open too.

13. <u>First solution of the $\mathcal{G}$-problem.</u>

<u>Lemma</u> 13.1. <u>Let</u> $U \in \mathcal{U}$, $K \in \mathcal{K}(X)$ <u>and</u> $\mu \in \mathcal{M}_+(X;t)$ <u>be given. Then there exists a</u> μ-<u>uniformity class</u> $\mathcal{F} \subseteq \mathcal{B}(X;[0,1])$ <u>and an open</u> μ-<u>continuous neighbourhood</u> N(K) <u>of</u> K <u>such that there to any</u> $f \in \mathcal{B}(X;[0,1])$ <u>can be found two functions</u> f^- <u>and</u> f^+ <u>fulfilling the following conditions</u>

(i): $f^- \in \mathcal{F}$, $f^+ \in \mathcal{F}$, $f^- \leq f^+$;

(ii): $f^- \leq f \leq f^+$ <u>on the set</u> N(K);

(iii): <u>For every</u> $\varepsilon > 0$ <u>the inequalities</u>

(13.1) $$\int_{N(K)} (f^+ - f)d\mu \leq \varepsilon \cdot \mu X + \mu_*(\partial_{u,\varepsilon}^+ f),$$

(13.2) $$\int_{N(K)} (f - f^-)d\mu \leq \varepsilon \cdot \mu X + \mu_*(\partial_{u,\varepsilon}^- f),$$

and

(13.3) $$\int (f^+ - f^-)d\mu \leq \varepsilon \cdot \mu X + \mu_*(\partial_{u,\varepsilon} f)$$

hold.

Proof. Choose $V \in \mathcal{U}$ so that $V \circ V^{-1} \subseteq U^{-1}$. Choose for each $x \in X$ an open μ-continuous neighbourhood $N(x)$ of x contained in $V[x]$ (see P16). We can find finitely many points $x_1,\ldots,x_n$ in K such that $K \subseteq \bigcup_1^n N(x_j)$. Put $A_1 = N(x_1)$ and, for $\nu = 2,\ldots,n$, put $A_\nu = N(x_\nu) \setminus \bigcup_1^{\nu-1} N(x_j)$. $A_1,\ldots,A_n$ are pairwise disjoint μ-continuity sets covering K. Furthermore, we have $A_j \subseteq V[x_j]$; $j=1,\ldots,n$. Denote by $\mathcal{F}$ the class of functions of the form $\Sigma_1^n \beta_j \cdot 1_{A_j}$ where $n \geq 1$ and $0 \leq \beta_j \leq 1$, $j=1,\ldots,n$. Clearly, $\mathcal{F}$ is a μ-uniformity class and the set $N(K) = \bigcup_1^n A_j = \bigcup_1^n N(x_j)$ is an open μ-continuous neighbourhood of K. Let $f \in \mathcal{B}(X;[0,1])$ be given. Define, for $j=1,\ldots,n$, β_j^- and β_j^+ by

$$\beta_j^- = \inf\{fx \mid x \in A_j\}, \quad \beta_j^+ = \sup\{fx \mid x \in A_j\},$$

and put $f^- = \Sigma_1^n \beta_j^- \cdot 1_{A_j}$, $f^+ = \Sigma_1^n \beta_j^+ \cdot 1_{A_j}$.

(13.1) now follows from the inequalities

$$\int_{N(K)} (f^+ - f)d\mu = \sum_j \left(\int_{A_j \cap \{\beta_j^+ - fx \leq \varepsilon\}} + \int_{A_j \cap \{\beta_j^+ - fx > \varepsilon\}} \right)$$
$$\leq \varepsilon \cdot \mu X + \sum_j \mu(A_j \cap \{\beta_j^+ - fx > \varepsilon\})$$
$$= \varepsilon \cdot \mu X + \mu\left(\bigcup_j A_j \cap \{\beta_j^+ - fx > \varepsilon\}\right)$$
$$\leq \varepsilon \cdot \mu X + \mu_*(\partial_{u,\varepsilon}^+ f).$$

The proof of (13.2) is similar and (13.3) follows from the inequalities

$$\int (f^+ - f^-) d\mu = \sum_j \omega_f(A_j) \cdot \mu(A_j)$$
$$\leq \varepsilon \cdot \mu X + \sum_{\{j \mid \omega_f A_j > \varepsilon\}} \omega_f A_j \cdot \mu A_j$$
$$\leq \varepsilon \cdot \mu X + \mu_* (\partial_{u,\varepsilon} f).$$

☐

In order to state our first result, we find it convinient to introduce the quantities $\eta\big((\mathcal{F}_i), \mu\big)$, $\eta^+\big((\mathcal{F}_i), \mu\big)$ and $\eta^-\big((\mathcal{F}_i), \mu\big)$ defined by

$$\eta = \sup_{\varepsilon > 0} \inf_{u \in \mathcal{U}} \limsup_i \sup_{f \in \mathcal{F}_i} \mu_*(\partial_{u,\varepsilon} f),$$

$$\eta^+ = \sup_{\varepsilon > 0} \inf_{u \in \mathcal{U}} \limsup_i \sup_{f \in \mathcal{F}_i} \mu_*(\partial^+_{u,\varepsilon} f),$$

$$\eta^- = \sup_{\varepsilon > 0} \inf_{u \in \mathcal{U}} \limsup_i \sup_{f \in \mathcal{F}_i} \mu_*(\partial^-_{u,\varepsilon} f).$$

These quantities are unchanged if we replace $\mathcal{U}$ by a base for $\mathcal{U}$. η is unchanged if we replace the sets $\partial_{U,\varepsilon} f$ occuring in the definition of η by their closures; this follows from (12.16). If we ignore some measurability problems, we see that the η-quantities are related by the inequalities $\max(\eta^+, \eta^-) \leq \eta \leq 2 \cdot \max(\eta^+, \eta^-)$.

Theorem 13.2. Let $(\mathcal{F}_i)_{i \in I}$ and μ be given and assume that $\mu \in \mathcal{M}_+(X; t)$. Then the inequalities $\xi^+ \leq \eta^+$, $\xi^- \leq \eta^-$ and $\xi \leq \eta$ hold, and the implications $\xi^+ = 0 \Leftrightarrow \eta^+ = 0$, $\xi^- = 0 \Leftrightarrow \eta^- = 0$ and $\xi = 0 \Leftrightarrow \eta = 0$ hold. Furthermore, in case each of the classes $\mathcal{F}_i$; $i \in I$ consists entirely of indicator-functions, then we have the more complete information that the equalities $\xi^+ = \eta^+$ and $\xi^- = \eta^-$ hold.

In the last mentioned case of the theorem we obtain $\xi = \max(\eta^+, \eta^-)$, and $\xi \leq \eta \leq 2\xi$ follows. Simple examples show that these ineqialities are best possible.

Proof. We begin by establishing the inequalities $\xi^+ \leq \eta^+$, $\xi^- \leq \eta^-$ and

$\xi \leq \eta$. Let $(\mu_\alpha, i_\alpha)_{\alpha\in D}$ be a subnet of $\mathfrak{N}(\mu)\times I$ and let, for some time, $U \in \mathcal{U}$ and $\varepsilon > 0$ be fixed. To a given $\tau > 0$ we find $K \in \mathcal{K}(X)$ such that $\mu(\complement K) < \tau$. Corresponding to U,K and μ we consider a class $\mathcal{F}$ and a set N(K) fulfilling the conditions of Lemma 11.1. Since $\mu_\alpha \xrightarrow{\mathcal{U}} \mu$, the ineqialities

$$\sup_{f\in\mathcal{F}} |\mu_\alpha(f) - \mu(f)| < \tau$$

and

$$\mu_\alpha(\complement N(K)) < \tau$$

hold eventually, say for $\alpha \geq \alpha_0$. Consider an $\alpha \geq \alpha_0$ and a function $f \in \mathcal{B}(X;[0,1])$. We have

$$\begin{aligned}\int f d\mu_\alpha - \int f d\mu &\leq \int f^+ d\mu_\alpha + \int_{\complement N(K)} f d\mu_\alpha - \int f d\mu \\ &\leq \left(\int f^+ d\mu_\alpha - \int f^+ d\mu\right) + \int f^+ d\mu - \int f d\mu + \tau \\ &\leq 2\tau + \int_{N(K)} (f^+ - f)\, d\mu + \int_{\complement N(K)} f^+ d\mu \\ &\leq 3\tau + \varepsilon\cdot\mu X + \mu(\partial^+_{U,\varepsilon} f).\end{aligned}$$

It then follows that

$$\begin{aligned}\limsup_\alpha \sup_{f\in\mathcal{F}_{i_\alpha}} (\mu_\alpha(f) - \mu(f)) &\leq \limsup_\alpha \sup_{f\in\mathcal{F}_{i_\alpha}} \mu(\partial^+_{U,\varepsilon} f) + \varepsilon\cdot\mu X \\ &\leq \limsup_i \sup_{f\in\mathcal{F}_i} \mu(\partial^+_{U,\varepsilon} f) + \varepsilon\cdot\mu X\end{aligned}$$

By the freedom of choice of $U \in \mathcal{U}$ we obtain

$$\begin{aligned}\limsup_\alpha \sup_{f\in\mathcal{F}_{i_\alpha}} (\mu_\alpha(f) - \mu(f)) &\leq \inf_{U\in\mathcal{U}} \limsup_i \sup_{f\in\mathcal{F}_i} \mu(\partial^+_{U,\varepsilon} f) + \varepsilon\cdot\mu X \\ &\leq \eta^+ + \varepsilon\cdot\mu X,\end{aligned}$$

and from this inequality we conclude that $\xi^+ \leq \eta^+$ holds. A similar argument shows that $\xi^- \leq \eta^-$ holds. The inequality $\xi \leq \eta$ follows from the two inequalities already established.

In order to establish the remaining parts of the theorem, assume that for some positive a, $\eta^+ > a$ holds. We can then find $\varepsilon > 0$ such that

$$\forall_{U \in \mathcal{U}} \; \forall_{i_0 \in I} \; \exists_{i \geq i_0} \; \exists_{f \in \mathcal{F}_i} \quad \mu_*(\partial^+_{U,\varepsilon} f) > a$$

Denote by D the set of those pairs $\alpha = (U,i)$ for which there exists $f \in \mathcal{F}_i$ with $\mu_*(\partial^+_{U,\varepsilon} f) > a$. When directing D in the natural way, we find that D is a subnet of $\mathcal{U} \times I$. Consider, for some time, a fixed element $\alpha = (U,i)$ in D and denote by f_α one of the functions in $\mathcal{F}_i$ with $\mu_*(\partial^+_{U,\varepsilon} f) > a$. Choose a compact subset K_α of $\partial^+_{U,\varepsilon}(f_\alpha)$ with $\mu(K_\alpha) > a$. It is easy to see that we can find finitely many pairwise disjoint Borel-sets $\{N_{\alpha,\nu}\}_{\nu=1,\dots,m(\alpha)}$ and finitely many points $\{x_{\alpha,\nu}\}_{\nu=1,\dots,m(\alpha)}$ in K_α such that

$$K_\alpha \subseteq \bigcup_1^{m(\alpha)} N_{\alpha,\nu} \tag{13.4}$$

and

$$x_{\alpha,\nu} \in N_{\alpha,\nu} \subseteq U[x_{\alpha,\nu}] \quad ; \quad \nu = 1,\dots,m(\alpha) \tag{13.5}$$

hold. We shall now construct a measure μ_α such that $\mu_\alpha(f_\alpha) - \mu(f_\alpha)$ is large and such that μ_α converges weakly to μ when α runs through D. μ_α will be constructed by selecting suitable points $\{y_{\alpha,\nu}\}_{\nu=1,\dots,m(\alpha)}$ and putting

$$\mu_\alpha = \mu \,\Big|\, \complement\left(\bigcup_1^{m(\alpha)} N_{\alpha,\nu}\right) + \sum_1^{m(\alpha)} \mu(N_{\alpha,\nu}) \cdot \varepsilon_{y_{\alpha,\nu}} , \tag{13.6}$$

where the first term is the restriction of μ to the set $\complement(\bigcup N_{\alpha,\nu})$. For any bounded measurable function f we have

$$\mu_\alpha(f) - \mu(f) = \sum_1^{m(\alpha)} \left(\mu(N_{\alpha,\nu}) \cdot f(y_{\alpha,\nu}) - \int_{N_{\alpha,\nu}} f \, d\mu \right). \tag{13.7}$$

Let us first look into the case where the classes $\mathcal{F}_i$ consists of indicator-functions. Let A_α denote the subset of X for which $f_\alpha = 1_{A_\alpha}$. In this case K_α is a subset of $\partial^+_U(A_\alpha) = U[A_\alpha] \setminus A_\alpha$. For each $\nu = 1,\dots,m(\alpha)$

we can choose $y_{\alpha,\nu}$ such that

$$y_{\alpha,\nu} \in \mathcal{U}^{-1}[x_{\alpha,\nu}] \cap A_\alpha \tag{13.8}$$

With this choice of the $y_{\alpha,\nu}$'s we find from (13.7) that

$$\begin{aligned}\mu_\alpha(A_\alpha) - \mu(A_\alpha) &= \textstyle\sum_1^{m(\alpha)} \{\mu(N_{\alpha,\nu}) - \mu(A_\alpha \cap N_{\alpha,\nu})\} \\ &= \mu\left(\textstyle\bigcup_1^{m(\alpha)} N_{\alpha,\nu} \setminus A_\alpha\right) \\ &\geq \mu(K_\alpha \setminus A_\alpha) \\ &= \mu K_\alpha \\ &> a.\end{aligned}$$

Since, by (13.8), the $y_{\alpha,\nu}$'s are close to the $x_{\alpha,\nu}$'s and since, by (13.5), the sets $N_{\alpha,\nu}$ are sets with small "diameter", it is not difficult to deduce from (13.7) that $\mu_\alpha(f) \to \mu(f)$ for any bounded function f, uniformly continuous w.r.t. $\mathcal{U}$. By Theorem 8.1 it follows that $\mu_\alpha \xrightarrow{w} \mu$. Since $(\mu_\alpha, i_\alpha)_{\alpha\in D}$, where $\alpha = (U_\alpha, i_\alpha)$, is a subnet of $\hat{N}(\mu)\times I$, it follows by the above calculations that $\xi^+ \geq a$. This argument proves the inequality $\eta^+ \leq \xi^+$. The inequality $\eta^- \leq \xi^-$ is proved analogously.

Now, let us turn to the general case where the functions in the classes $\mathcal{F}_1$ need not be indicator-functions. In that case, we first choose a fixed positive number ε' such that

$$\varepsilon' < \min\left(\frac{\varepsilon}{2}, \frac{\varepsilon \cdot a}{4 \cdot \mu X}\right). \tag{13.9}$$

We then choose, for each $\nu = 1, \ldots, m(\alpha)$, the point $y_{\alpha,\nu}$ such that the two requirements

$$y_{\alpha,\nu} \in U^{-1} \circ U[x_{\alpha,\nu}]$$

and

$$f_\alpha(y_{\alpha,\nu}) > \sup\{f_\alpha(y) \mid y \in \mathcal{U}^{-1} \circ \mathcal{U}[x_{\alpha,\nu}]\} - \varepsilon'$$

are fulfilled. We claim, that then

(13.10) $$\sup\{f_\alpha(y) \mid y \in K_\alpha \cap N_{\alpha,\nu}\} \leq f_\alpha(y_{\alpha,\nu}) - \frac{\varepsilon}{2} \quad ; \quad \nu = 1,\dots,m(\alpha)$$

holds. To see this, assume that $y_o \in K_\alpha \cap N_{\alpha,\nu}$. From the definition of $\partial^+_{U,\varepsilon}$ and from the inclusion $K_\alpha \subseteq \partial^+_{U,\varepsilon}(f_\alpha)$ we find $z \in U^{-1}[y_o]$ with $f_\alpha(z) > f_\alpha(y_o)+\varepsilon$. Since $y_o \in N_{\alpha,\nu} \subseteq U[x_{\alpha,\nu}]$ we have

$$\begin{aligned} f_\alpha(y_o) &< f_\alpha(z) - \varepsilon \leq \sup\{f_\alpha(y) \mid y \in U^{-1}\circ U[x_{\alpha,\nu}]\} - \varepsilon \\ &< f_\alpha(y_{\alpha,\nu}) + \varepsilon' - \varepsilon \leq f_\alpha(y_{\alpha,\nu}) - \frac{\varepsilon}{2} \quad , \end{aligned}$$

and (13.10) follows. Consider the measure μ_α defined by (13.6). We have

$$\begin{aligned} \mu_\alpha(f_\alpha) - \mu(f_\alpha) &= \sum_\nu \Big\{\mu(N_{\alpha,\nu}\cap K_\alpha)\cdot f_\alpha(y_{\alpha,\nu}) - \int_{N_{\alpha,\nu}\cap K_\alpha} f_\alpha\, d\mu\Big\} \\ &\quad + \sum_\nu \Big\{\mu(N_{\alpha,\nu}\setminus K_\alpha)\cdot f_\alpha(y_{\alpha,\nu}) - \int_{N_{\alpha,\nu}\setminus K_\alpha} f_\alpha\, d\mu\Big\} \\ &\geq \sum_\nu \frac{\varepsilon}{2}\mu(N_{\alpha,\nu}\cap K_\alpha) - \sum_\nu \varepsilon'\cdot\mu(N_{\alpha,\nu}\setminus K_\alpha) \\ &\geq \frac{1}{2}\varepsilon\cdot a - \varepsilon'\cdot\mu X \\ &\geq \frac{1}{4}\varepsilon\cdot a . \end{aligned}$$

From this inequality it follows in a similar manner as before that $\xi^+ \geq \frac{1}{4}\varepsilon\cdot a$. In particular, we have $\xi^+ > 0$. We have seen that $\eta^+>0 \Rightarrow \xi^+>0$. Taken together with the inequality $\xi^+ \leq \eta^+$, we obtain the desired result: $\xi^+ = 0 \Leftrightarrow \eta^+ = 0$. The equivalence $\xi^- = 0 \Leftrightarrow \eta^- = 0$ is proved in an analogous way. The same can be said about the equivalence $\xi=0 \Leftrightarrow \eta=0$, or we may use the inequality $\eta \leq 2\cdot\max(\eta^+,\eta^-)$. ▯

14. <u>Second solution of the ξ-problem.</u> We now introduce a quantity $\zeta = \zeta((\mathcal{F}_i),\mu)$ defined by

$$\zeta = \sup_{\varepsilon>0} \sup_{(U_\alpha, i_\alpha, f_\alpha)} \mu_*\Big(\bigcap_\alpha \partial_{U_\alpha,\varepsilon} f_\alpha\Big) ,$$

where the second supremum is taken over all nets $(U_\alpha, i_\alpha, f_\alpha)_{\alpha\in D}$ on

$\mathcal{U} \times I \times \mathcal{B}(X;[0,1]))$ such that $(U_\alpha, i_\alpha)_{\alpha \in D}$ is a subnet of $\mathcal{U} \times I$ and such that, for each $\alpha \in D$, $f_\alpha \in \mathcal{F}_{i_\alpha}$.

From (12.16) it follows that the sets $\partial_{U_\alpha,\varepsilon}(f_\alpha)$ occuring in the definition of ζ may be replaced by their closures, thus it is not necessary to work with sets that are not known to be measurable.

We shall prove that $\zeta = \eta$ holds. The inequality $\zeta \leq \eta$ is rather obvious. The inequality $\eta \leq \zeta$ is less obvious and it is this inequality that is the more useful one. The inclusion (12.15) is the key fact needed to estblish the desired result. This beeing so, we prove a lemma of a more general nature.

If $\mathcal{E} \subseteq \mathfrak{D}(X)$ and $U \in \mathcal{U}$, we denote by $U[\mathcal{E}]$ the class of U-sections of sets in $\mathcal{E}$:

$$U[\mathcal{E}] = \{U[E] \mid E \in \mathcal{E}\}.$$

If $\mathcal{E}'$ is another class of subsets of X, we write $\mathcal{E} < \mathcal{E}'$ if, to any set E in $\mathcal{E}$, there exists a set E' in $\mathcal{E}'$ with $E \subseteq E'$.

Lemma 14.1. Let I be a directed set and assume that to each pair $(U,i) \in \mathcal{U} \times I$ there is given a class $\mathcal{E}_{U,i} \subseteq \mathfrak{D}(X)$. Assume that the two requirements

$$U \subseteq V \Rightarrow \mathcal{E}_{U,i} < \mathcal{E}_{V,i}$$

and

$$U[\mathcal{E}_{V,i}] < \mathcal{E}_{U \circ V,i}$$

are fulfilled $(U,V \in \mathcal{U},\ i \in I)$. Furthermore, a Radon measure $\mu \in \mathcal{M}_+(X;t)$ is given. Define η by

$$\eta = \inf_{U \in \mathcal{U}} \limsup_{i} \sup_{E \in \mathcal{E}_{U,i}} \mu_*(E),$$

and define ζ by

$$\zeta = \sup_{(U_\alpha, i_\alpha, E_\alpha)} \mu_*\Big(\bigcap_\alpha E_\alpha\Big).$$

where the supremum is taken over all nets $(U_\alpha, i_\alpha, E_\alpha)_{\alpha\in D}$ on $\mathcal{U}\times I\times \mathfrak{D}(X)$ such that $(U_\alpha, i_\alpha)_{\alpha\in D}$ is a subnet of $\mathcal{U}\times I$ and such that $E_\alpha \in \mathcal{E}_{U_\alpha, i_\alpha}$ for each $\alpha \in D$.

Then $\eta = \zeta$.

Proof. The simple proof of the inequality $\zeta \leq \eta$ is left to the reader.

To prove the remaining inequality, assume that $\eta > a$ for some $a \geq 0$. From the definition of η it follows that there exists a positive ε and a net $(U_\alpha, i_\alpha, E_\alpha)_{\alpha\in D}$ of the type appearing in the definition of ζ such that

$$\mu_*(E_\alpha) > a+\varepsilon; \quad \alpha \in D$$

holds. Choose to each $\alpha \in D$ a closed subset F_α of E_α such that $\mu F_\alpha > a+\varepsilon$ holds. A subnet of $(F_\alpha)_{\alpha\in D}$ converges in the notion of convergence discussed in P . We may assume that the net $(F_\alpha)_{\alpha\in D}$ itself converges, say $F_\alpha \to F$. Since

$$F = \bigcap_\alpha \overline{\bigcup_{\beta\geq\alpha} F_\beta},$$

and since μ is τ-smooth, we find that $\mu F \geq a+\varepsilon$. Now choose a compact subset K of F for which $\mu K \geq a$ holds. Consider a set $U \in \mathcal{U}$. We can find finitely many points $\{x_\nu\}_{\nu=1,\dots,n}$ in K such that

$$K \subseteq \bigcup_1^n U[x_\nu]$$

holds. For each $\nu=1,\dots,n$ we have $U[x_\nu]\cap F_\alpha \neq \emptyset$, eventually. Combining these facts, we have $K \subseteq U\circ U^{-1}[F_\alpha]$, eventually. We have

$$\mathcal{U}\circ\mathcal{U}^{-1}[F_\alpha] \subseteq \mathcal{U}\circ\mathcal{U}^{-1}[E_\alpha] \in \mathcal{U}\circ\mathcal{U}^{-1}[\mathcal{E}_{\mathcal{U}_\alpha, i_\alpha}] < \mathcal{E}_{\mathcal{U}\circ\mathcal{U}^{-1}\circ\mathcal{U}_\alpha, i_\alpha}$$

What we have seen can be expressed shortly as follows

$$\forall_{\mathcal{U}} \ \exists_{\alpha_0} \ \forall_{\alpha\geq\alpha_0} \ \exists_{E\in\mathcal{E}_{\mathcal{U}\circ\mathcal{U}^{-1}\circ\mathcal{U}_\alpha, i_\alpha}} \quad K \subseteq E$$

Bearing in mind that $\mu K \geq a$, it is not difficult to see from this that $\zeta \geq a$ must hold. ▯

It follows from the lemma that $\eta\big((\mathcal{F}_i),\mu\big) = \zeta\big((\mathcal{F}_i),\mu\big)$ in case $\mu \in \mathcal{M}_+(X;t)$. From Theorem 13.2 we then obtain

Theorem 14.2. Let $(\mathcal{F}_i)_{i\in I}$ and $\mu \in \mathcal{M}_+(X;t)$ be given. Then $\xi = 0$ holds if and only if $\zeta = 0$ holds.

Let us here give one application of this result.

Theorem 14.3. Let $(\mathcal{F}_i)_{i\in I}$ be given. A necessary and sufficient condition that

$$\xi\big((\mathcal{F}_i),\mu\big) = 0$$

for every $\mu \in \mathcal{M}_+(X;t)$ is that the following condition is fulfilled:

(14.1)
$$\forall_{x\in X} \ \forall_{(x_\alpha, i_\alpha)} \ \lim_\alpha \sup_{f\in\mathcal{F}_{i_\alpha}} |f(x_\alpha) - f(x)| = 0 \ ;$$

here "$\forall(x_\alpha, i_\alpha)$" is short for "for every net $(x_\alpha, i_\alpha)_{\alpha\in D}$ such that $x_\alpha \to x$ and such that $(i_\alpha)_{\alpha\in D}$ is a subnet of I".

Note that the condition (14.1) is equivalent to the condition that $\xi\big((\mathcal{F}_i),\varepsilon_x\big)$ be equal to 0 for every $x \in X$ provided we change the definition of ξ only paying attention to measures in $\mathcal{E} = \{\varepsilon_x | x\in X\}$.

Proof. Necessity follows from the above remark.

It follows from Theorem 14.2 that the necessary and sufficient condition for ξ to be 0 for every $\mu \in \mathcal{M}_+(X;t)$ is that the condition

$$\forall_{(U_\alpha, i_\alpha, f_\alpha)} \ \bigcup_{\varepsilon>0} \ \bigcap_\alpha \partial_{U_\alpha,\varepsilon} f_\alpha = \emptyset$$

holds $\big((U_\alpha, i_\alpha)$ a subnet of $\mathcal{U} \times I$, $f_\alpha \in \mathcal{F}_{i_\alpha}$; $\alpha \in D\big)$. If this condition does not hold, there exists a point $x \in X$, a net $(U_\alpha, i_\alpha, f_\alpha)$ of the proper type, and a positive ε such that, for each α, $x \in \partial_{U_\alpha,\varepsilon}(f_\alpha)$. To each α we can then find a point x_α in $U_\alpha^{-1}[x]$ such that

$$|f_\alpha(x_\alpha) - f_\alpha(x)| > \varepsilon/2.$$

Since $x_\alpha \to x$ and since (i_α) is a subnet of I, (14.1) does not hold. ▯

15. Uniformity classes. In this and in the next two sections we discuss some applications of the results obtained in sections 13 and 14.

We first examine the case when I contains only one point i. What we have given is then a class $\mathcal{F} = \mathcal{F}_i \subseteq \mathcal{B}(X;[0,1])$ and a tight measure μ. To say that $\xi = 0$ is the same as saying that $\mathcal{F}$ is a μ-uniformity class. From Theorem 14.2 and the remark following the definition of ζ we see that $\xi = 0$ holds if and only if

$$(15.1) \qquad \forall_{\varepsilon>0} \ \forall_{(f_u)} \ \mu\Big(\bigcap_{u\in\mathcal{U}} \partial_{u,\varepsilon}(f_u)\Big) = 0$$

holds, where "$\forall(f_u)$" is short for "for every net $(f_u)_{u\in\mathcal{U}}$ of functions in $\mathcal{F}$ indexed by $\mathcal{U}$". Condition (15.1) is equivalent with the condition

$$(15.2) \qquad \forall_{(f_u)} \ \mu\Big(\bigcup_{\varepsilon>0} \bigcap_{u\in\mathcal{U}} \overline{\partial_{u,\varepsilon}(f_u)}\Big) = 0 ,$$

and also with the condition

$$(15.3) \qquad \forall_{(f_u)} \ \mu_\bullet\Big(\bigcup_{\varepsilon>0} \bigcap_{u\in\mathcal{U}} \partial_{u,\varepsilon}(f_u)\Big) = 0$$

$\mathcal{F}$ is a μ-uniformity class for every $\mu \in \mathcal{M}_+(X;t)$ if and only if

$$\forall_{(f_u)} \ \bigcup_{\varepsilon>0} \bigcap_{u\in\mathcal{U}} \partial_{u,\varepsilon}(f_u) = \emptyset$$

holds, and then also if and only if

$$\bigcup_{\varepsilon>0} \bigcap_{u\in\mathcal{U}} \bigcup_{f\in\mathcal{F}} \partial_{u,\varepsilon} f = \emptyset$$

holds. In other words, $\mathcal{F}$ is a μ-uniformity class for every $\mu \in \mathcal{M}_+(X;t)$ if and only if $\mathcal{F}$ is everywhere equicontinuous (this result can of course also be derived from Theorem 14.3).

If $\mathcal{F}$ consists of only one function f, we see from (15.3) that $\mu_\alpha(f) \to \mu(f)$ for every net (μ_α) converging weakly to μ if and only if f is a μ-continuity function.

If all functions in $\mathcal{F}$ are indicator-functions, say $\mathcal{F} = \{1_A | A \in \mathcal{A}\}$ then (15.3) takes the following form:

$$(15.4) \qquad \forall_{(A_u)} \quad \mu_* \Big(\bigcap_{u \in \mathcal{U}} \partial_u (A_u) \Big) = 0 .$$

From (15.4) we derive the powerful result, that if $\mathcal{A}$ satisfies the condition

$$(15.5) \qquad \forall_{(A_u) \subseteq \mathcal{A}} \quad \exists_{A \in \mathcal{A}} \quad \bigcap_{u \in \mathcal{U}} \partial_u (A_u) \subseteq \partial A$$

and if $\mathcal{A}$ is a μ-continuity class (i.e. $\mu(\partial A) = 0 \ \forall A \in \mathcal{A}$) then $\mathcal{A}$ is a μ-uniformity class.

We shall now prove that if $\mathcal{A}$ satisfies condition (15.5) and if every set in $\mathcal{A}$ is either open or closed (or, more generally, if to any $A \in \mathcal{A}$ there exists a set $A' \subseteq \mathcal{A}$ such that $\partial A \subseteq \partial A'$ holds and such that A' is either closed or open) then

$$(15.6) \qquad \xi(\mathcal{A}, \mu) = \sup_{A \in \mathcal{A}} \mu(\partial A)$$

for any $\mu \in \mathcal{M}_+(X;t)$. Before the proof, we remark that this result may for instance be applied to the class of convex measurable subsets of an Euclidean space.

To prove the result, first remark that we have

$$\xi = \xi^+ \vee \xi^- = \eta^+ \vee \eta^- \leq \eta = \zeta = \sup_{A \in \mathcal{A}} \mu(\partial A).$$

Then observe that we have:

$$\begin{aligned} \eta^+ \vee \eta^- &= \max\Big(\inf_u \sup_A \mu_*(\partial_u^+ A) , \inf_u \sup_A \mu_*(\partial_u^- A) \Big) \\ &\geq \max\Big(\sup_A \inf_u \mu_*(\partial_u^+ A) , \sup_A \inf_u \mu_*(\partial_u^- A) \Big) \\ &\geq \max\Big(\sup_A \mu(\partial A \cap \complement A), \sup_A \mu(\partial A \cap A) \Big) \end{aligned}$$

$$\geq \sup_A \max\left(\mu(\partial A \cap \complement A), \mu(\partial A \cap A)\right)$$

Under the stated conditions it follows from this that $\eta^+ \vee \eta^- \geq \zeta \cdot \xi = \zeta$ follows.

16. Joint continuity. We pose the problem: Is the map $(\mu,f) \to \mu(f)$ jointly continuous? In the space of measures we insist on the weak topology and in $\hat{R}$ we of course insist on the usual topology. The problem is thus really this: Can we define a suitable topology, or, since we are not so demanding, a suitable notion of convergence, in the space of functions such that $(\mu,f) \to \mu(f)$ is continuous? Since we want every constant net of functions to converge, we have to restrict our attention to continuous functions. We choose only to consider the class $C(X;[0,1])$ of continuous functions $f: X \to [0,1]$.

Let $\xrightarrow{c}$ be a notion of convergence on $C(X;[0,1])$. We say that $\xrightarrow{c}$ is compatible with the mapping (duality) $(\mu,f) \to \mu(f)$, or, in more detail, compatible with the mapping $\mathcal{M}_+(X;t) \times C(X;[0,1]) \to \hat{R}$, if we have $\mu_\alpha(f_\alpha) \to \mu(f)$ whenever $\mu_\alpha \xrightarrow{w} \mu$ and $f_\alpha \xrightarrow{c} f$. The notion $\xrightarrow{c}$ is said to be compatible with the mapping $(f,x) \to f(x)$, or, in more detail, compatible with the mapping $C(X;[0,1]) \times X \to \hat{R}$, if we have $f_\alpha(x_\alpha) \to f(x)$ whenever $f_\alpha \xrightarrow{c} f$ and $x_\alpha \to x$.

It is only natural to quess that the topology on $C(X;[0,1])$ of uniform convergence on compact sets emerges in the problem of joint continuity; we denote convergence in this topology by the symbol $\xrightarrow{u.c.}$. For convinience, we mention the following elementary result:

Lemma 16.1. For functions in $C(X;[0,1])$ we have $f_i \xrightarrow{u.c.} f$ if and only if for every net $(x_\alpha, i_\alpha)_{\alpha \in D}$ on $X \times I$ such that (i_α) is a subnet of I, such that $\{x_\alpha | \alpha \in D\}$ is relatively compact, and such that, for some $x \in X$, $x_\alpha \to x$ we have $f_\alpha(x_\alpha) \to f(x)$.

This lemma implies that if a notion of convergence is compatible with $(f,x) \to f(x)$ then the implication $f_\alpha \to_t f \Rightarrow f_\alpha \xrightarrow{u.c.} f$ holds.

Theorem 16.2. A notion of convergence on $C(X;[0,1])$ is compatible with $(\mu,f) \to \mu(f)$ if and only if it is compatible with $(f,x) \to f(x)$.

Proof. The "only if" part is trivial. Assume that $\to_t$ is compatible with $(f,x) \to f(x)$. Let $(f_i)_{i\in I}$ be a convergent net in the notion $\to_t$, say $f_i \to_t f$. Put, for $i \in I$, $\mathcal{F}_i = \{f_i\}$ and consider the ξ-problem associated with $(\mathcal{F}_i)_{i\in I}$. We shall prove that condition (14.1) is satisfied; therefore, we consider $x \in X$ and a net (x_α, i_α) of the type appearing in (14.1). Since $f_{i_\alpha}(x_\alpha) \to f(x)$ and $f_{i_\alpha}(x) \to f(x)$ we have $f_{i_\alpha}(x_\alpha) - f_{i_\alpha}(x) \to 0$. Thus, condition (14.1) is satisfied. If $(\mu_i)_{i\in I}$ is a convergent net on $\mathcal{M}_+(X;t)$, say $\mu_i \xrightarrow{w} \mu$, we can now conclude that

$$\lim_i |\mu_i(f_i) - \mu(f_i)| = 0.$$

Since $f_i \xrightarrow{u.c.} f$ we have $\mu(f_i) \to \mu(f)$, and $\mu_i(f_i) \to \mu(f)$ follows. ∎

In the reminder of this section $\to_t$ denotes a specific notion of convergence on $C(X;[0,1])$, called the notion of continuous convergence, and defined by taking $f_i \to_t f$ to mean that, for every x and $(x_\alpha, i_\alpha)_{\alpha\in D}$ such that $x_\alpha \to x$ and such that $(i_\alpha)_{\alpha\in D}$ is a subnet of I, we have $f_{i_\alpha}(x_\alpha) \to f(x)$.

It is easy to see that $\to_t$ is in fact a notion of convergence. By Theorem 16.2, it follows that $\to_t$ is the weakest notion of convergence on $C(X;[0,1])$ compatible with $(\mu,f) \to \mu(f)$.

It is not difficult to prove that $f_i \to_t f$ holds if and only if the condition

(16.1) $$\forall_{x} \ \forall_{\varepsilon>0} \ \exists_{N(x)} \ \| (f_i - f)|_{N(x)} \| < \varepsilon, \text{ eventually}$$

is satisfied ($N(x)$ denotes a neighbourhood of x). Condition (16.1) is equivalent with the condition

$$(16.2) \qquad \forall_{K\in\mathcal{K}(X)} \ \forall_{\varepsilon>0} \ \exists_{N(K)} \ \| (f_i - f)|_{N(K)} \| < \varepsilon \text{, eventually.}$$

If X is first countable and $(f_n)_{n\geq 1}$ a sequence on $C(X;[0,1])$, we have

$$(16.3) \qquad f_n \to_c f \Leftrightarrow f_n \xrightarrow{u.c.} f.$$

This can easily be derived from (16.1).

It is fairly easy to see that uniform convergence on compact sets is compatible with $(\mu,f) \to \mu(f)$ if and only if X is locally compact. It is harder to obtain the following result.

Theorem 16.3. The notion of continuous convergence is a topological notion of convergence if and only if X is locally compact.

This follows from Theorem 3 of Arens [2].

17. Preservation of weak convergence. Let X and Y denote two completely regular spaces, $\mathcal{U}$ a uniformity for X and $\mathcal{V}$ a uniformity for Y. The image of a measure $\mu \in \mathcal{M}_+(X)$ under a measurable mapping $h: X \to Y$ is the measure $h\circ\mu$, sometimes also denoted μh^{-1}, in $\mathcal{M}_+(Y)$ defined by

$$h\circ\mu\ (A) = \mu\big(h^{-1}(A)\big); \quad A \in \mathcal{B}(Y).$$

Let $(h_i)_{i\in I}$ be a net of measurable mappings $X \to Y$, μ a measure in $\mathcal{M}_+(X)$ and ν a measure in $\mathcal{M}_+(Y)$. We shall say that weak convergence is preserved if, for any subnet $(\mu_\alpha, i_\alpha)_{\alpha\in D}$ of $\hat{N}(\mu)\times I$ it is true that $h_{i_\alpha}\circ\mu_\alpha \xrightarrow{w} \nu$. From now on we assume that $\mu \in \mathcal{M}_+(X;t)$ and $\nu \in \mathcal{M}_+(Y;\tau)$. From Theorem 8.1 we see that a necessary and sufficient condition that weak convergence be preserved is that

$$(17.1) \qquad \forall_{g\in C(Y;[0,1])} \ \forall_{(\mu_\alpha, i_\alpha)} \ \int g(h_{i_\alpha})\, d\mu_\alpha \to \int g\, d\nu$$

holds (as usual, (μ_α, i_α) is a subnet of $\hat{N}(\mu)\times I$). It is easy to see that (17.1) is equivalent to the two requirements:

$$(17.2) \qquad h_i\circ\mu \xrightarrow{w} \nu ,$$

(17.3) $$\forall_{g \in C(Y;[0,1])} \quad \xi\left(\left(\{g(h_i)\}\right)_{i\in I}, \mu\right) = 0.$$

Since we have assumed that μ is tight, the results of sections 12 and 13 apply.

For $U \in \mathcal{U}$, $V \in \mathcal{V}$ and $h: X \to Y$ we denote by $\partial_{U,V}(h)$ the set

$$\partial_{U,V}(h) = \{x \mid \exists_{x',x'' \in U^{-1}[x]} (hx',hx'') \notin V\}.$$

__Theorem__ 17.1. __Assuming that__ $\mu \in \mathcal{M}_+(X;t)$ __and that__ $\nu \in \mathcal{M}_+(Y;\tau)$, __we have that weak convergence is preserved if and only if__

$$h_i \circ \mu \xrightarrow{w} \nu$$

__and__

(17.4) $$\forall_{V \in \mathcal{V}} \quad \inf_{U \in \mathcal{U}} \limsup_i \mu_*(\partial_{U,V} h_i) = 0$$

__hold. The last condition is equivalent to the condition__

(17.5) $$\forall_{V \in \mathcal{V}} \quad \forall_{(U_\alpha,i_\alpha)} \quad \mu_*\left(\bigcap_\alpha \partial_{U_\alpha,V}(h_{i_\alpha})\right) = 0,$$

__where__ $(U_\alpha,i_\alpha)_{\alpha\in D}$ __denotes a subnet of__ $\mathcal{U}\times I$.

__Proof__. To prove sufficiency, assume that (17.2) and (17.4) hold. We have to prove that (17.3) holds. It is enough to prove that

$$\xi\left((\{g\circ h_i\})_{i\in I}, \mu\right) = 0$$

holds for any $g: Y \to [0,1]$ which is uniformly continuous w.r.t. $\mathcal{V}$. When g is uniformly continuous, we can, to any $\varepsilon > 0$, find $V_\varepsilon \in \mathcal{V}$ such that

$$(y',y'') \in V_\varepsilon \Rightarrow |g(y')-g(y'')| < \varepsilon$$

It follows, that the inclusion

$$\partial_{U,\varepsilon}(g\circ h) \subseteq \partial_{U,V_\varepsilon}(h)$$

holds, and, by (17.4), we find that

$$\eta\Big((\{g\circ h_i\})_{i\in I},\ \mu\Big) = 0.$$

Theorem 13.2 then implies the desired result.

To prove necessity, assume that weak convergence is preserved. Then (17.2) holds. We shall first prove that (17.4) holds for a specific uniformity $\mathcal{V}^*$ on Y. $\mathcal{V}^*$ is defined by taking as a subbase all sets of the form

$$\{(y',y'')\,\big|\ |g(y')-g(y'')| < \varepsilon\ \}$$

where $g \in C(Y;[0,1])$ and ε is positive. Assume that $V \in \mathcal{V}^*$. Then there exists finitely many functions $g_1,\dots,g_n$ in $C(Y;[0,1])$ and $\varepsilon > 0$ such that

$$V \supseteq \cap_1^n\{(y',y'')\,\big|\ |g_\nu(y')-g_\nu(y'')| \leq \varepsilon\}.$$

It follows that for any h: X $\to$ Y we have

$$\partial_{U,V}(h) \subseteq \cup_1^n\ \partial_{U,\varepsilon}(g_\nu\circ h).$$

Now it is not difficult to establish (17.4), applying (17.3) and Theorem 13.2.

Still assuming that weak convergence is preserved, we want to establish (17.4) for any uniformity $\mathcal{V}$ on Y (compatible with the topology of Y). For $V^* \in \mathcal{V}^*$ and $V \in \mathcal{V}$ we denote by $A(V^*,V)$ the subset of Y defined by

$$A(V^*,V) = \{y\,|\ V^*[y] \nsubseteq V[y]\}.$$

It is a matter of straight-forward considerations to check that, for each fixed $V \in \mathcal{V}$, the class of sets $\overline{A(V^*,V)}$; $V^* \in \mathcal{V}^*$ filters downwards to the empty set:

(17.6) $$\forall_{V\in\mathcal{V}}\ \ \overline{A(V^*,V)} \downarrow \emptyset.$$

We need another elementary fact:

(17.7) $$\partial_{U,V^{-1}\circ V}(h) \subseteq h^{-1}\big(A(V^*,V)\big) \cup \partial_{U,V^*}(h).$$

Consider a fixed V in $\mathcal{V}$. Since (17.4) is known to hold when we use $\mathcal{V}^*$

instead of $\mathcal{V}$, we obtain from (17.7) that, for any $V^* \in \mathcal{V}^*$, we have

$$\inf_{U\in\mathcal{U}} \limsup_i \mu_*\Big(\partial_{U,V}(h_i)\Big) \leq \limsup_i \mu h_i^{-1}\Big(\overline{A(V^*,V)}\Big)$$

$$\leq \nu\Big(\overline{A(V^*,V)}\Big).$$

Employing (17.6) and the τ-smoothness of ν we deduce from this that

$$\inf_{U\in\mathcal{U}} \limsup_i \mu_*(\partial_{U,V}h_i) = 0$$

holds. ▯

N O T E S   A N D   R E M A R K S

PRELIMINARIES

For the result of P7 see Bourbaki [6], chap.I, §10, ex. 1.

The results of P8 can be found in Effros [8].

P11 can for instance be proved by applying the well known idea in the proof of 6.B of Halmos [9].

Many of the elementary properties of τ-smooth and tight measures can be found in Varadarajan [27].

P16. The fact that a τ-smooth measure on a locally compact space is tight, follows from the observation that the paving $\mathcal{G}$ of open, relatively compact sets filters to the right to X (also employ the regularity of μ). The corresponding fact for a space provided with a complete metric is proved by an "$\varepsilon\cdot 2^{-n}$-argument" where, for each n, we work with the paving $\mathcal{G}_n$ consisting of all sets that are finite unions of open spheres with radius 1/n; to get through along these lines, we just have to recall that a totally bounded subset of a complete metric space is relatively compact.

The first result of P19 is taken from Topsøe [24]. This result may be generalized, considering a class of functions instead of a class of sets, to the following result: Let $\mathcal{F}$ be a lattice of functions containing the two constant functions 0 and 1 and separating points T_2 in the sense that, for every pair (x,y) of distinct points in X, there exists a function f in $\mathcal{F}$ with $f_*(x) = 1$ and $f^*(y) = 0$; if μ_1 and μ_2 are tight, and if $\mu_1(f_*) \leq \mu_2(f^*)$ for all $f \in \mathcal{F}$, then $\mu_1 \leq \mu_2$ holds.

SECTION 2

Theorem 2.2 for X a topological space and $\mathcal{K} = \mathcal{K}(X)$ is due to Kisyński (cf. Theorem 1.2 of [14]). For a topological space, $\mathcal{G}(\mathcal{K}(X))$ is the class of open sets forthe associated k-space. Thus, if X is a k-space, for instance if X is first-countable or locally compact, we have $\mathcal{G}(\mathcal{K}(X)) = \mathcal{G}(X)$. We do not know if it is always true that $\mathcal{B}(\mathcal{K}(X)) = \mathcal{B}(X)$.

The proof of Lemma 2.4, (ii) is taken from Kisyński (cf. Theorem 1.1 of [14]).

SECTION 3

Lemma 3.5 is due to Ditlev Monrad. This lemma allowed us to drop the extra assumption that $k\backslash 1 \in \mathcal{K}$ for all $k \in \mathcal{K}^{\cdot}$, an assumption which was imposed at an ealier stage of the work when a lemma analogous to Lemma 4.7 played an important role.

Note that if $I_*(1) < \infty$ holds then the considerations following Lemma 3.5 can be simplified.

Lemma 3.7 was pointed out to us by N. Holger Petersen.

Comparing with Lemma 2.1, it is natural to conjecture that if $\mathcal{K}$ and λ satisfy the assumptions of Theorem 3.13 and if λ is τ-smooth at 0 then λ will be τ-smooth (and hence I will be τ-smooth w.r.t. $\mathcal{U}_+(\mathcal{K})$). We have proved this conjecture if we further assume that λ is locally finite (this notion beeing defined in analogy with the corresponding notion from section 2).

SECTION 4

Theorem 1,§7 is due to A. Markoff (cf. Theorem of [1]).

SECTION 5

As already mentioned, I am indebted to E.T. Kehlet for a helpful discussion. The actual arrangement of the steps (i)-(vi) in the proof of Theorem 5.1 is due to Tue Tjur (an ealier version worked with an extra assumption of local finiteness).

A similar result as Theorem 5.1 may be derived by taking as starting point a $(\emptyset,\cup f,\cap f)$-paving $\mathcal{K}_o$ and a tight and σ-smooth set-function λ_o on $\mathcal{K}_o$; then $\mathcal{K}$ should denote the $(\emptyset,\cup f,\cap c)$-closure of $\mathcal{K}_o$ and λ the set-function on $\mathcal{K}$ defined by

$$\lambda K = \inf_{K_o \supseteq K} \lambda_o K_o ;\ K \in \mathcal{K} .$$

Returning to Theorem 5.1 as it stands, note that, since $\mathcal{K}$ is a $(\emptyset,\cup f,\cap a)$-paving, $\mathcal{G}(\mathcal{K})$ defines a topology on X, and note that μ is a τ-smooth measure in this space (here, we better allow the topological space to be non-Hausdorff). Thus we have seen that "all abstract τ-smooth measure theory is topological". Had we assumed that $\mathcal{K}_o$ were a compact paving, we would find that "all abstract compact measure theory is topological"; for this to be true, we should allow non-Hausdorff topological spaces and then base the "compact" or "tight" measure theory on set-functions defined on the compact paving consisting of all closed and compact subsets.

SECTION 6

Those results of this section depending on set-functions defined on $\mathcal{G}(X)$ can be generalized by replacing $\mathcal{G}(X)$ with a $(\emptyset,\cup f,\cap f)$-paving of open sets separating points.

In the approach of Halmos, Theorem 6.2, (ii) plays an important role (cf. 53.E of [9]).

SECTION 7

The investigations of this section were stimulated by discussions with E. Alfsen.

Note that condition (iv) of Theorem 7.1 is fulfilled if X is completely regular and if the mappings $\mu \to \mu(f)$ with f bounded and continuous are lower semi-continuous.

Lemma 7.5 is a simple consequence of Theorem 2.5.2 in Michael [16], however, the proof in the text is very direct.

In establishing Lemma 7.7 we have been influenced by LeCam, who employed the idea behind the proof of (i) to establish a particular case of Theorem 9.3 (cf. Theorem 4 of [15]).

If X does not have property (*), $\mathcal{P}_+(X;t)$ need not be sequentially Hausdorff in the topology ϑ; this follows by inspection of Varadarajans example, p.225 of [27].

SECTION 8

The definition of the weak topology (or, more correctly, the topology of weak convergence) is, in a sense, new. However, due to previous research (Alexandroff [1] and Varadarajan [27]), it is a very natural definition. The definition is expressed in terms of semi-continuity, but may as well be expressed in terms of continuity, provided we change the topology of the real line.

We have chosen to work entirely in a topological set-up in part II, even though a more axiomatic setting is possible (cf. [26]). Recall, that the notes and remarks to section 5 implies that it is very likely that we will find ourselves in a topological setting anyhow.

The proof of Theorem 8.1 does not reveal anything new.

SECTION 9

The proof cf. (iv), Theorem 9.1 resembles closely that given in the notes and remarks to P16.

To my mind, the most interesting unsolved problem on compactness in the weak topology is the problem to characterize those spaces X for which the relatively compact subsets of $\mathcal{M}_+(X;t)$ and the tight subsets of $\mathcal{M}_+(X;t)$ (with μX bounded) are the same. In [27] Varadarajan claimes that every metrizable space has this property; unfortunately, his proof only works in the locally compact case (see the remark preceeding Theorem 16.3). Even if one considers the space of rationals, it does not seem to be known if the desired property holds; the best result we have been able to obtain in this special case is the following: If $\mathcal{P}$ is a compact subset of the set of measures on the rationals, and if ε is positive then there exists a closed, totally bounded and nowhere dense set F such that $\mu(\complement F) < \varepsilon$ for every $\mu \in \mathcal{P}$.

SECTION 10

Some of the results have been taken from [24].

The second condition in Theorem 10.1 may of course be replaced by the condition

$$\liminf \mu_\alpha \bar{A} \geq \mu \mathring{A}; \quad A \in \mathcal{A} .$$

Theorem 10.1 may be generalized to a class of functions, see the notes and remarks to P19.

Theorem 10.4 is, essentially, due to Varadarajan (see Theorem II.5 of [27]). In case X is regular, we may replace the main condition of Theorem 10.4 by the condition

$$\liminf \mu_\alpha \bar{A} \geq \mu \mathring{A}; \quad A \in \mathcal{A} ,$$

since, in that case, $\mathcal{A}$ will separate points and closed sets T_2.

Corollary 10.5 is a convenient generalization of Theorem 2.2 of Billingsley [3]. It is, for instance, easy to obtain the usual characterization of weak convergence on Euclidean spaces from this corollary.

SECTION 11

Theorem 11.1 is partially, if not entirely, known.

In Theorem 11.2, (ii) and (v) are known (see Varadarajan [27]) and, what we only observed recently, (iii) is known too (see Blau [5]). For similar properties see Kallianpur [12] and Varadarajan [27]. Note that the result that $\mathcal{M}_+(X;\tau)$ is separable and metrizable if and only if X is so follows from (iii) and (iv). One could add to Theorem 11.2 the result already proved that $\mathcal{M}_+^1(X;t)$ (or $\mathcal{M}_+^1(X;\tau)$) is compact if and only if X is so.

SECTIONS 12,13,14

The results given generalize those of Billingsley and Topsøe [4] and Topsøe [22] and, furthermore, contain a result announced in Topsøe [23].

The reader should observe that in our main results we assume that the measure μ is tight. This is indeed a very convenient assumption, but it is not necessary to make it (τ-smoothness will do, but it is not entirely trivial to see this).

The paper [25] contains some applications of the results of section 13 to Glivenko-Cantelli problems.

SECTION 17

Compare with the results of [23].

R E F E R E N C E S

[1] Alexandroff, A.D.: Additive set functions in abstract spaces. Mat. Sb.8,307-348(1940); 9,563-628(1941); 13,169-238(1943).

[2] Arens, R.: A topology for spaces of transformations. Ann.of Math. (2)47,480-495(1946).

[3] Billingsley, P.: Convergence of probability measures. New York: Wiley and Sons 1968.

[4] Billingsley, P., and Topsøe, F.: Uniformity in weak convergence. Z . Wahrscheinlichkeitstheorie verw.Geb. 7,1-16 (1967).

[5] Blau, J.H.: The space of measures on a given set. Fundamenta Mathematicae 38,23-34(1951).

[6] Bourbaki, N.: Éléments de mathématique, Livre III, Topologie générale, 2ed. Paris: Hermann et Cie 1951.

[7] Carathéodory, C.: Über das lineare Mass von Punktmengen - eine Verallgemeinerung des Längenbegriffs. Nachr.Akad.Wiss. Göttingen Math.-Phys.Kl.II,404-426(1914).

[8] Effros, E.G.: Convergence of closed subsets in a topological space. Proc.Amer.Math.Soc. 16,929-931(1965).

[9] Halmos, P.R.: Measure theory. New York: D.Van Nostrand 1950.

[10] Hausdorff, F.: Über halbstetige Funktionen und deren Verallgemeinerung, Mathematische Zeitschrift 5,292-309(1915).

[11] Tong, H.: Some characterizations of normal and perfectly normal spaces, Duke Math.J.19,289-292(1952).

[12] Kallianpur, G.: The topology of weak convergence of probability measures. J.Math.Mech.10,947-969(1961)

[13] Kelley, J.L.: General topology. New York: D. Van Nostrand 1955.

[14] Kisyński, J.: On the generation of tight measures, Studia Math.30, 141-151(1968)

[15] LeCam, L.: Convergence in distribution of stochastic processes. Univ.California Publ.Statist.2, no.11,207-236(1957).

[16] Michael, E.: Topologies on spaces of subsets, Trans.Amer.Math.Soc. 371,152-182(1951).

[17] Parthasarathy, K.R.: Probability measures on metric spaces. New York: Acad.Press 1967.

[18] Pettis, B.J.: On the extension of measures. Ann.of Math.54,186-197(1951).

[19] Prohorov, Yu .V.: Convergence of random processes and limit theorems in probability theory. Theor.Prob.Appl.1,157-214(1956).

[20] Schwartz, L.: Radon measures on Sousliu spaces. Queen's papers in pure and applied mathematics, no.10,157-168(1967).

[21] Srinivasan, T.P.: On extensions of measures, J.of the Indian Math. Soc.19,31-60(1955).

[22] Topsøe, F.: On the connection between P-continuity and P-uniformity in weak convergence. Theor.Prob.Appl.12,281-290(1967).

[23] Topsøe, F.: Preservation of weak convergence under mappings. Ann. Math.Statist.38,1661-1665(1967).

[24] Topsøe, F.: A criterion for weak convergence of measures with an application to convergence of measures on D[0,1]. Math.Scand., to appear.

[25] Topsøe, F.: On the Glivenko-Cantelli theorem. Z. Wahrscheinlichkeitstheorie verw.Geb.14,239-250(1970).

[26] Topsøe, F.: Compactness in spaces of measures. Studia Math., to appear.

[27] Varadarajan, V.S.: Measures on topological spaces. Amer.Math.Soc. Transl.ser.II,48,161-228(1965).

[28] Weil, A.: Sur les espaces à structure uniforme et sur la topologie générale. Paris: Actualités Sci.Ind.551(1937).

Offsetdruck: Julius Beltz, Weinheim/Bergstr

Lecture Notes in Mathematics

Bisher erschienen/Already published

Vol. 1: J. Wermer, Seminar über Funktionen-Algebren. IV, 30 Seiten. 1964. DM 3,80 / $ 1.10

Vol. 2: A. Borel, Cohomologie des espaces localement compacts d'après. J. Leray. IV, 93 pages. 1964. DM 9,– / $ 2.60

Vol. 3: J. F. Adams, Stable Homotopy Theory. Third edition IV, 78 pages. 1969. DM 8,– / $ 2.20

Vol. 4: M. Arkowitz and C. R. Curjel, Groups of Homotopy Classes. 2nd. revised edition. IV, 36 pages. 1967. DM 4,80 / $ 1.40

Vol. 5: J.-P. Serre, Cohomologie Galoisienne Troisième édition. VIII, 214 pages. 1965. DM 18,– / $ 5.00

Vol. 6: H. Hermes, Eine Termlogik mit Auswahloperator. IV, 42 Seiten. 1965. DM 5,80 / $ 1.60

Vol. 7: Ph. Tondeur, Introduction to Lie Groups and Transformation Groups. Second edition. VIII, 176 pages. 1969. DM 14,– / $ 3.80

Vol. 8: G. Fichera, Linear Elliptic Differential Systems and Eigenvalue Problems. IV, 176 pages. 1965. DM 13,50 / $ 3.80

Vol. 9: P. L. Ivănescu, Pseudo-Boolean Programming and Applications. IV, 50 pages. 1965. DM 4,80 / $ 1.40

Vol. 10: H. Lüneburg, Die Suzukigruppen und ihre Geometrien. VI, 111 Seiten. 1965. DM 8,– / $ 2.20

Vol. 11: J.-P. Serre, Algèbre Locale. Multiplicités. Rédigé par P. Gabriel. Seconde édition. VIII, 192 pages. 1965. DM 12,– / $ 3.30

Vol. 12: A. Dold, Halbexakte Homotopiefunktoren. II, 157 Seiten. 1966. DM 12,– / $ 3.30

Vol. 13: E. Thomas, Seminar on Fiber Spaces. IV, 45 pages 1966. DM 4,80 / $ 1.40

Vol. 14: H. Werner, Vorlesung über Approximationstheorie. IV, 184 Seiten und 12 Seiten Anhang. 1966. DM 14,– / $ 3.90

Vol. 15: F. Oort, Commutative Group Schemes. VI, 133 pages. 1966. DM 9,80 / $ 2.70

Vol. 16: J. Pfanzagl and W. Pierlo, Compact Systems of Sets. IV, 48 pages. 1966. DM 5,80 / $ 1.60

Vol. 17: C. Müller, Spherical Harmonics. IV, 46 pages. 1966. DM 5,– / $ 1.40

Vol. 18: H.-B. Brinkmann und D. Puppe, Kategorien und Funktoren. XII, 107 Seiten, 1966. DM 8,– / $ 2.20

Vol. 19: G. Stolzenberg, Volumes, Limits and Extensions of Analytic Varieties. IV, 45 pages. 1966. DM 5,40 / $ 1.50

Vol. 20: R. Hartshorne, Residues and Duality. VIII, 423 pages. 1966. DM 20,– / $ 5.50

Vol. 21: Seminar on Complex Multiplication. By A. Borel, S. Chowla, C. S. Herz, K. Iwasawa, J.-P. Serre. IV, 102 pages. 1966. DM 8,– / $ 2.20

Vol. 22: H. Bauer, Harmonische Räume und ihre Potentialtheorie. IV, 175 Seiten. 1966. DM 14,– / $ 3.90

Vol. 23: P. L. Ivănescu and S. Rudeanu, Pseudo-Boolean Methods for Bivalent Programming. 120 pages. 1966. DM 10,– / $ 2.80

Vol. 24: J. Lambek, Completions of Categories. IV, 69 pages. 1966. DM 6,80 / $ 1.90

Vol. 25: R. Narasimhan, Introduction to the Theory of Analytic Spaces. IV, 143 pages. 1966. DM 10,– / $ 2.80

Vol. 26: P.-A. Meyer, Processus de Markov. IV, 190 pages. 1967. DM 15,– / $ 4.20

Vol. 27: H. P. Künzi und S. T. Tan, Lineare Optimierung großer Systeme. VI, 121 Seiten. 1966. DM 12,– / $ 3.30

Vol. 28: P. E. Conner and E. E. Floyd, The Relation of Cobordism to K-Theories. VIII, 112 pages. 1966. DM 9,80 / $ 2.70

Vol. 29: K. Chandrasekharan, Einführung in die Analytische Zahlentheorie. VI, 199 Seiten. 1966. DM 16,80 / $ 4.70

Vol. 30: A. Frölicher and W. Bucher, Calculus in Vector Spaces without Norm. X, 146 pages. 1966. DM 12,– / $ 3.30

Vol. 31: Symposium on Probability Methods in Analysis. Chairman. D. A. Kappos. IV, 329 pages. 1967. DM 20,– / $ 5.50

Vol. 32: M. André, Méthode Simpliciale en Algèbre Homologique et Algèbre Commutative. IV, 122 pages. 1967. DM 12,– / $ 3.30

Vol. 33: G. I. Targonski, Seminar on Functional Operators and Equations. IV, 110 pages. 1967. DM 10,– / $ 2.80

Vol. 34: G. E. Bredon, Equivariant Cohomology Theories. VI, 64 pages. 1967. DM 6,80 / $ 1.90

Vol. 35: N. P. Bhatia and G. P. Szegö, Dynamical Systems. Stability Theory and Applications. VI, 416 pages. 1967. DM 24,– / $ 6.60

Vol. 36: A. Borel, Topics in the Homology Theory of Fibre Bundles. VI, 95 pages. 1967. DM 9,– / $ 2.50

Vol. 37: R. B. Jensen, Modelle der Mengenlehre. X, 176 Seiten. 1967. DM 14,– / $ 3.90

Vol. 38: R. Berger, R. Kiehl, E. Kunz und H.-J. Nastold, Differentialrechnung in der analytischen Geometrie IV, 134 Seiten. 1967 DM 12,– / $ 3.30

Vol. 39: Séminaire de Probabilités I. II, 189 pages. 1967. DM 14,– / $ 3.90

Vol. 40: J. Tits, Tabellen zu den einfachen Lie Gruppen und ihren Darstellungen. VI, 53 Seiten. 1967. DM 6.80 / $ 1.90

Vol. 41: A. Grothendieck, Local Cohomology. VI, 106 pages. 1967. DM 10,– / $ 2.80

Vol. 42: J. F. Berglund and K. H. Hofmann, Compact Semitopological Semigroups and Weakly Almost Periodic Functions. VI, 160 pages. 1967. DM 12,– / $ 3.30

Vol. 43: D. G. Quillen, Homotopical Algebra. VI, 157 pages. 1967. DM 14,– / $ 3.90

Vol. 44: K. Urbanik, Lectures on Prediction Theory. IV, 50 pages. 1967. DM 5,80 / $ 1.60

Vol. 45: A. Wilansky, Topics in Functional Analysis. VI, 102 pages. 1967. DM 9,60 / $ 2.70

Vol. 46: P. E. Conner, Seminar on Periodic Maps. IV, 116 pages. 1967. DM 10,60 / $ 3.00

Vol. 47: Reports of the Midwest Category Seminar I. IV, 181 pages. 1967. DM 14,80 / $ 4.10

Vol. 48: G. de Rham, S. Maumary et M. A. Kervaire, Torsion et Type Simple d'Homotopie. IV, 101 pages. 1967. DM 9,60 / $ 2.70

Vol. 49: C. Faith, Lectures on Injective Modules and Quotient Rings. XVI, 140 pages. 1967. DM 12,80 / $ 3.60

Vol. 50: L. Zalcman, Analytic Capacity and Rational Approximation. VI, 155 pages. 1968. DM 13.20 / $ 3.70

Vol. 51: Séminaire de Probabilités II IV, 199 pages. 1968. DM 14,– / $ 3.90

Vol. 52: D. J. Simms, Lie Groups and Quantum Mechanics. IV, 90 pages. 1968. DM 8,– / $ 2.20

Vol. 53: J. Cerf, Sur les difféomorphismes de la sphère de dimension trois ($\Gamma_4 = 0$). XII, 133 pages. 1968. DM 12,– / $ 3.30

Vol. 54: G. Shimura, Automorphic Functions and Number Theory. VI, 69 pages. 1968. DM 8,– / $ 2.20

Vol. 55: D. Gromoll, W. Klingenberg und W. Meyer, Riemannsche Geometrie im Großen. VI, 287 Seiten. 1968. DM 20,– / $ 5.50

Vol. 56: K. Floret und J. Wloka, Einführung in die Theorie der lokalkonvexen Räume. VIII, 194 Seiten. 1968. DM 16,– / $ 4.40

Vol. 57: F. Hirzebruch und K. H. Mayer, O (n)-Mannigfaltigkeiten, exotische Sphären und Singularitäten. IV, 132 Seiten. 1968. DM 10,80 / $ 3.00

Vol. 58: Kuramochi Boundaries of Riemann Surfaces. IV, 102 pages. 1968 DM 9,60 / $ 2.70

Vol. 59: K. Jänich, Differenzierbare G-Mannigfaltigkeiten. VI, 89 Seiten. 1968. DM 8,– / $ 2.20

Vol. 60: Seminar on Differential Equations and Dynamical Systems. Edited by G. S. Jones. VI, 106 pages. 1968. DM 9,60 / $ 2.70

Vol. 61: Reports of the Midwest Category Seminar II. IV, 91 pages. 1968. DM 9,60 / $ 2.70

Vol. 62: Harish-Chandra, Automorphic Forms on Semisimple Lie Groups X, 138 pages. 1968. DM 14,– / $ 3.90

Vol. 63: F. Albrecht, Topics in Control Theory. IV, 65 pages. 1968. DM 6,80 / $ 1.90

Vol. 64: H. Berens, Interpolationsmethoden zur Behandlung von Approximationsprozessen auf Banachräumen. VI, 90 Seiten. 1968. DM 8,– / $ 2.20

Vol. 65: D. Kölzow, Differentiation von Maßen. XII, 102 Seiten. 1968. DM 8,– / $ 2.20

Vol. 66: D. Ferus, Totale Absolutkrümmung in Differentialgeometrie und -topologie. VI, 85 Seiten. 1968. DM 8,– / $ 2.20

Vol. 67: F. Kamber and P. Tondeur, Flat Manifolds. IV, 53 pages. 1968. DM 5,80 / $ 1.60

Vol. 68: N. Boboc et P. Mustață, Espaces harmoniques associés aux opérateurs différentiels linéaires du second ordre de type elliptique. VI, 95 pages. 1968. DM 8,60 / $ 2.40

Vol. 69: Seminar über Potentialtheorie. Herausgegeben von H. Bauer. VI, 180 Seiten. 1968. DM 14,80 / $ 4.10

Vol. 70: Proceedings of the Summer School in Logic. Edited by M. H. Löb. IV, 331 pages. 1968. DM 20,– / $ 5.50

Vol. 71: Séminaire Pierre Lelong (Analyse), Année 1967 – 1968. VI, 19 pages. 1968. DM 14,– / $ 3.90

Bitte wenden / Continued

Vol. 72: The Syntax and Semantics of Infinitary Languages. Edited by J. Barwise. IV, 268 pages. 1968. DM 18,– / $ 5.00

Vol. 73: P. E. Conner, Lectures on the Action of a Finite Group. IV, 123 pages. 1968. DM 10,– / $ 2.80

Vol. 74: A. Fröhlich, Formal Groups. IV, 140 pages. 1968. DM 12,– / $ 3.30

Vol. 75: G. Lumer, Algèbres de fonctions et espaces de Hardy. VI, 80 pages. 1968. DM 8,– / $ 2.20

Vol. 76: R. G. Swan, Algebraic K-Theory. IV, 262 pages. 1968. DM 18,– / $ 5.00

Vol. 77: P.-A. Meyer, Processus de Markov: la frontière de Martin. IV, 123 pages. 1968. DM 10,– / $ 2.80

Vol. 78: H. Herrlich, Topologische Reflexionen und Coreflexionen. XVI, 166 Seiten. 1968. DM 12,– / $ 3.30

Vol. 79: A. Grothendieck, Catégories Cofibrées Additives et Complexe Cotangent Relatif. IV, 167 pages. 1968. DM 12,– / $ 3.30

Vol. 80: Seminar on Triples and Categorical Homology Theory. Edited by B. Eckmann. IV, 398 pages. 1969. DM 20,– / $ 5.50

Vol. 81: J.-P. Eckmann et M. Guenin, Méthodes Algébriques en Mécanique Statistique. VI, 131 pages. 1969. DM 12,– / $ 3.30

Vol. 82: J. Wloka, Grundräume und verallgemeinerte Funktionen. VIII, 131 Seiten. 1969. DM 12,– / $ 3.30

Vol. 83: O. Zariski, An Introduction to the Theory of Algebraic Surfaces. IV, 100 pages. 1969. DM 8,– / $ 2.20

Vol. 84: H. Lüneburg, Transitive Erweiterungen endlicher Permutationsgruppen. IV, 119 Seiten. 1969. DM 10,– / $ 2.80

Vol. 85: P. Cartier et D. Foata, Problèmes combinatoires de commutation et réarrangements. IV, 88 pages. 1969. DM 8,– / $ 2.20

Vol. 86: Category Theory, Homology Theory and their Applications I. Edited by P. Hilton. VI, 216 pages. 1969. DM 16,– / $ 4.40

Vol. 87: M. Tierney, Categorical Constructions in Stable Homotopy Theory. IV, 65 pages. 1969. DM 6,– / $ 1.70

Vol. 88: Séminaire de Probabilités III. IV, 229 pages. 1969. DM 18,– / $ 5.00

Vol. 89: Probability and Information Theory. Edited by M. Behara, K. Krickeberg and J. Wolfowitz. IV, 256 pages. 1969. DM 18,– / $ 5.00

Vol. 90: N. P. Bhatia and O. Hajek, Local Semi-Dynamical Systems. II, 157 pages. 1969. DM 14,– / $ 3.90

Vol. 91: N. N. Janenko, Die Zwischenschrittmethode zur Lösung mehrdimensionaler Probleme der mathematischen Physik. VIII, 194 Seiten. 1969. DM 16,80 / $ 4.70

Vol. 92: Category Theory, Homology Theory and their Applications II. Edited by P. Hilton. V, 308 pages. 1969. DM 20,– / $ 5.50

Vol. 93: K. R. Parthasarathy, Multipliers on Locally Compact Groups. III, 54 pages. 1969. DM 5,60 / $ 1.60

Vol. 94: M. Machover and J. Hirschfeld, Lectures on Non-Standard Analysis. VI, 79 pages. 1969. DM 6,– / $ 1.70

Vol. 95: A. S. Troelstra, Principles of Intuitionism. II, 111 pages. 1969. DM 10,– / $ 2.80

Vol. 96: H.-B. Brinkmann und D. Puppe, Abelsche und exakte Kategorien, Korrespondenzen. V, 141 Seiten. 1969. DM 10,– / $ 2.80

Vol. 97: S. O. Chase and M. E. Sweedler, Hopf Algebras and Galois theory. II, 133 pages. 1969. DM 10,– / $ 2.80

Vol. 98: M. Heins, Hardy Classes on Riemann Surfaces. III, 106 pages. 1969. DM 10,– / $ 2.80

Vol. 99: Category Theory, Homology Theory and their Applications III. Edited by P. Hilton. IV, 489 pages. 1969. DM 24,– / $ 6.60

Vol. 100: M. Artin and B. Mazur, Etale Homotopy. II, 196 Seiten. 1969. DM 12,– / $ 3.30

Vol. 101: G. P. Szegö et G. Treccani, Semigruppi di Trasformazioni Multivoche. VI, 177 pages. 1969. DM 14,– / $ 3.90

Vol. 102: F. Stummel, Rand- und Eigenwertaufgaben in Sobolewschen Räumen. VIII, 386 Seiten. 1969. DM 20,– / $ 5.50

Vol. 103: Lectures in Modern Analysis and Applications I. Edited by C. T. Taam. VII, 162 pages. 1969. DM 12,– / $ 3.30

Vol. 104: G. H. Pimbley, Jr., Eigenfunction Branches of Nonlinear Operators and their Bifurcations. II, 128 pages. 1969. DM 10,– / $ 2.80

Vol. 105: R. Larsen, The Multiplier Problem. VII, 284 pages. 1969. DM 18,– / $ 5.00

Vol. 106: Reports of the Midwest Category Seminar III. Edited by S. Mac Lane. III, 247 pages. 1969. DM 16,– / $ 4.40

Vol. 107: A. Peyerimhoff, Lectures on Summability. III, 111 pages. 1969. DM 8,– / $ 2.20

Vol. 108: Algebraic K-Theory and its Geometric Applications. Edited by R. M. F. Moss and C. B. Thomas. IV, 86 pages. 1969. DM 6,– / $ 1.70

Vol. 109: Conference on the Numerical Solution of Differential Equations. Edited by J. Ll. Morris. VI, 275 pages. 1969. DM 18,– / $ 5.00

Vol. 110: The Many Facets of Graph Theory. Edited by G. Chartrand and S. F. Kapoor. VIII, 290 pages. 1969. DM 18,– / $ 5.00

Vol. 111: K. H. Mayer, Relationen zwischen charakteristischen Zahlen. III, 99 Seiten. 1969. DM 8,– / $ 2.20

Vol. 112: Colloquium on Methods of Optimization. Edited by N. N. Moiseev. IV, 293 pages. 1970. DM 18,– / $ 5.00

Vol. 113: R. Wille, Kongruenzklassengeometrien. III, 99 Seiten. 1970. DM 8,– / $ 2.20

Vol. 114: H. Jacquet and R. P. Langlands, Automorphic Forms on GL (2). VII, 548 pages. 1970. DM 24,– / $ 6.60

Vol. 115: K. H. Roggenkamp and V. Huber-Dyson, Lattices over Orders I. XIX, 290 pages. 1970. DM 18,– / $ 5.00

Vol. 116: Séminaire Pierre Lelong (Analyse) Année 1969. IV, 195 pages. 1970. DM 14,– / $ 3.90

Vol. 117: Y. Meyer, Nombres de Pisot, Nombres de Salem et Analyse Harmonique. 63 pages. 1970. DM 6.– / $ 1.70

Vol. 118: Proceedings of the 15th Scandinavian Congress, Oslo 1968. Edited by K. E. Aubert and W. Ljunggren. IV, 162 pages. 1970. DM 12,– / $ 3.30

Vol. 119: M. Raynaud, Faisceaux amples sur les schémas en groupes et les espaces homogènes. III, 219 pages. 1970. DM 14,– / $ 3.90

Vol. 120: D. Siefkes, Büchi's Monadic Second Order Successor Arithmetic. XII, 130 Seiten. 1970. DM 12,– / $ 3.30

Vol. 121: H. S. Bear, Lectures on Gleason Parts. III, 47 pages. 1970. DM 6,– / $ 1.70

Vol. 122: H. Zieschang, E. Vogt und H.-D. Coldewey, Flächen und ebene diskontinuierliche Gruppen. VIII, 203 Seiten. 1970. DM 16,– / $ 4.40

Vol. 123: A. V. Jategaonkar, Left Principal Ideal Rings. VI, 145 pages. 1970. DM 12,– / $ 3.30

Vol. 124: Séminare de Probabilités IV. Edited by P. A. Meyer. IV, 282 pages. 1970. DM 20,– / $ 5.50

Vol. 125: Symposium on Automatic Demonstration. V, 310 pages. 1970. DM 20,– / $ 5.50

Vol. 126: P. Schapira, Théorie des Hyperfonctions. XI, 157 pages. 1970. DM 14,– / $ 3.90

Vol. 127: I. Stewart, Lie Algebras. IV, 97 pages. 1970. DM 10,– / $ 2.80

Vol. 128: M. Takesaki, Tomita's Theory of Modular Hilbert Algebras and its Applications. II, 123 pages. 1970. DM 10,– / $ 2.80

Vol. 129: K. H. Hofmann, The Duality of Compact Semigroups and C*-Bigebras. XII, 142 pages. 1970. DM 14,– / $ 3.90

Vol. 130: F. Lorenz, Quadratische Formen über Körpern. II, 77 Seiten. 1970. DM 8,– / $ 2.20

Vol. 131: A Borel et al., Seminar on Algebraic Groups and Related Finite Groups. VII, 321 pages. 1970. DM 22,– / $ 6.10

Vol. 132: Symposium on Optimization. III, 348 pages. 1970. DM 22,– / $ 6.10

Vol. 133: F. Topsøe, Topology and Measure. XIV, 79 pages. 1970. DM 8,– / $ 2.20

MIX
Papier aus verantwortungsvollen Quellen
Paper from responsible sources
FSC® C105338

If you have any concerns about our products,
you can contact us on
ProductSafety@springernature.com

In case Publisher is established outside the EU,
the EU authorized representative is:
Springer Nature Customer Service Center GmbH
Europaplatz 3, 69115 Heidelberg, Germany

Printed by Libri Plureos GmbH
in Hamburg, Germany